AA007315

XII All Russian Conference "Thermophysics and Power Engineering in Academic Centers" (TPEAC 2021)

Journal of Physics: Conference Series Volume 2150

Sochi, Russia
25 – 27 October 2021

ISBN: 978-1-7138-5121-9
ISSN: 1742-6588

Printed from e-media with permission by:

Curran Associates, Inc.
57 Morehouse Lane
Red Hook, NY 12571

Some format issues inherent in the e-media version may also appear in this print version.

This work is licensed under a Creative Commons Attribution 3.0 International Licence.
Licence details: http://creativecommons.org/licenses/by/3.0/.

No changes have been made to the content of these proceedings. There may be changes to pagination and minor adjustments for aesthetics.

Printed with permission by Curran Associates, Inc. (2023)

For permission requests, please contact the Institute of Physics at the address below.

Institute of Physics
Dirac House, Temple Back
Bristol BS1 6BE UK

Phone: 44 1 17 929 7481
Fax: 44 1 17 920 0979

techtracking@iop.org

Additional copies of this publication are available from:

Curran Associates, Inc.
57 Morehouse Lane
Red Hook, NY 12571 USA
Phone: 845-758-0400
Fax: 845-758-2633
Email: curran@proceedings.com
Web: www.proceedings.com

TABLE OF CONTENTS

Journal of Physics: Conference Series **2150** (2022) 011001 doi:10.1088/1742-6596/2150/1/011001

PREFACE

the XII All-Russian Conference "Thermophysics and Power Engineering in Academic Centers" (TPEAC 2021)

October 25-27, 2021, Sochi, Russia

The All-Russian scientific conference "Thermophysics and Power Engineering in Academic Centers" (TPEAC-2021) continues series of conferences (workshops) on thermophysics and heat power engineering. Since the first conference until now, the principal organizer of the conference is the Kutateladze Institute of Thermophysics, SB RAS. The 1st Workshop of the University Departments and Research Teams in Siberia and the far East specializing in thermophysics was held in Novosibirsk in 1978. Since then, the geographical spread of the Conference expanded every year involving various research centers: Tomsk, 1980; Krasnoyarsk, 1982; Vladivostok, 1984; Kemerovo, 1986; Ulan-Ude, 1988. The 7th Workshop was held in Novosibirsk, September 9-12, 1990, at the Institute of Thermophysics SB RAS. Workshop of Universities in Siberia and Far East on Thermophysics and Heat Power Engineering, dedicated to the 85th Birthday of the Workshop founder, academician S.S. Kutateladze, was held in Novosibirsk, October 6-8, 1999, at the Institute of Thermophysics SB RAS, and the Novosibirsk State Technical University (NSTU). Since 1999 new series of workshops started. The 2nd Workshop was held in Tomsk, 2001, at the Tomsk Polytechnic University (TPU); 3rd - in Barnaul, 2003, at the Altai State Technical University; 4th - in Vladivostok, 2005, at the Far-East State Technical University; 5th - in Irkutsk, 2007, at the Irkutsk State Technical University. Since 2009, when the 6th Workshop was held in Krasnoyarsk at the Siberian Federal University, it acquired the status of an All-Russian conference. The 7th All-Russian Workshop of Universities on Thermophysics and Power Engineering was held in Kemerovo, 2011, at the Kuzbas State Technical University; 8th - in Yekaterinburg, 2013, at the Ural Federal University. Since 2015, when the 9th Workshop was held in Kazan, at the Kazan State Engineering University, it acquired the status of an international conference. International Conference "The 10th Workshop of Universities on Thermophysics and Power Engineering", was held in Moscow, 2017, at the Moscow Engineering University. The 11th All-Russian scientific conference with international participation "Thermophysics and Power Engineering in Academic Centers" (TPEAC-2019) held based on the Peter the Great St. Petersburg Polytechnic University (SPbPU), Saint Petersburg.

The Conference TPEAC-2021 was traditionally attended by students, graduate students, specialists, both young and leading scientists of universities, academic institutes, research centers, and other organizations. The Conference is dedicated to discussing the latest results and promising research areas in a wide range of fundamental and applied scientific problems of thermophysics, thermal power engineering, nuclear energy, and several related problems, including environmental safety. The Scientific Committee considers the Conference as an important contribution to the further development of energy technologies and energy equipment, and many other technical devices with the decisive role of heat and mass transfer processes.

This volume of Journal of Physics, Conf. Series contains 35 papers that were presented at the 12th All-Russian scientific conference “Thermophysics and Power Engineering in Academic Centers” (TPEAC-2021) held in October 25-27, 2021, Sochi, Russia. These papers were peer-reviewed by the Scientific Committee of the Conference and selected for publication in this volume.

Editorial Committee

- S.V. Alekseenko
- D.M. Markovich
- P.A. Kuibin

Organizing Committee

Chairman

Content from this work may be used under the terms of the Creative Commons Attribution 3.0 licence. Any further distribution of this work must maintain attribution to the author(s) and the title of the work, journal citation and DOI.
Published under licence by IOP Publishing Ltd

Journal of Physics: Conference Series **2150** (2022) 011001 doi:10.1088/1742-6596/2150/1/011001

- S. V. Alekseenko

Deputy Chairman

- E.N. Kalaidin
- P.A. Kuibin

Scientific secretaries

- S.G. Skripkin

Scientific Committee

Chairman

- D.M. Markovich

Deputy Chairman

- V.A Pershukov
- E.N. Kalaidin
- P.A. Kuibin

Members of the Scientific and Advisory Committee

- T.F. Bogatova
- A.R. Bogomolov
- L.A. Bolshov
- A.V. Dedov
- A.A. Dekterev
- A.S. Dmitriev
- S.M. Dmitriev
- S.L. Elistratov
- A.S. Zavorin
- V.I. Ilgisonis
- V.A. Isaev
- S.A. Isaev
- V.V. Klimenko
- Yu.A. Kuzma – Kichta
- V.V. Kuznetsov
- A.L. Kupershtokh
- O.O. Milman
- O.V. Mitrofanova
- A.V. Mityakov
- I.V. Naumov
- V.V. Nikonenko
- A.N. Pavlenko
- Yu.K. Petrenya
- V.V. Petrunin
- B.G. Pokusaev
- I.A. Popov
- V.I. Pachkov
- A.F. Ryzhkov
- S.Z. Sapozhnikov
- V.A. Stennikov

- E.V. Stroganova
- V.I. Terekhov
- A.N. Tygov
- M.P. Fyodorov
- E.D. Fedorovich
- V.Yu. Khomich
- N.D. Chichirova
- K.A. Shtym
- P.A. Shchinnikov

Peer review declaration

All papers published in this volume of Journal of Physics: Conference Series have been peer reviewed through processes administered by the Editors. Reviews were conducted by expert referees to the professional and scientific standards expected of a proceedings journal published by IOP Publishing.

- **Type of peer review: Single-blind / Double-blind / Triple-blind / Open / Other (please describe)**
- **Conference submission management system:**
- **Number of submissions received: 35**
- **Number of submissions sent for review: 35**
- **Number of submissions accepted: 34**
- **Acceptance Rate (Number of Submissions Accepted / Number of Submissions Received X 100): 97**
- **Average number of reviews per paper: 2**
- **Total number of reviewers involved: 33**
- **Any additional info on review process:**
- **Contact person for queries: Alexandr Kvon tpeac2021@gmail.com**

Content from this work may be used under the terms of the Creative Commons Attribution 3.0 licence. Any further distribution of this work must maintain attribution to the author(s) and the title of the work, journal citation and DOI.
Published under licence by IOP Publishing Ltd

Comparative analysis of air and water flows in simplified hydraulic turbine models

S G Skripkin[1,2*], D A Suslov[1,2], I V Litvinov[1,2], E U Gorelikov[1,2], M A Tsoy[1,2] and S I Shtork[1,2]

[1] Kutateladze Institute of Thermophysics, Novosibirsk, 630090, Russia
[2] Novosibirsk State University, Novosibirsk, 630090, Russia

skripkin.s.g@gmail.com

Abstract. This article presents a comparative analysis of flow characteristics behind a hydraulic turbine runner in air and water. Swirling flow with a precessing vortex core (PVC) was investigated using a laser Doppler anemometer and pressure pulsation sensors. The experiments were conducted on aerodynamic and hydrodynamic test rigs over a wide range of hydraulic turbine operating conditions. Part-load modes of hydraulic turbine operation were investigated using the Fourier transform of pressure pulsations obtained from acoustic sensors. The features of the swirling flow were shown for the range of operating conditions from deep partl-load to overload.

1. Introduction

Wind, solar, and thermal power plants are very slow in terms of power regulation. Hydropower plants provide stable operation of electricity grid. The regulation of hydro turbine operation based on the amount of energy required leads to operation in non-optimal modes in which the flow coming off the edges of the water turbine runner is highly swirled and highly susceptible to disturbances. The expanding part of the hydraulic turbine draft tube leads to the formation of a powerful precessing vortex core (PVC) of spiral shape in the flow. The vortex rope causes significant periodic pressure pulsations in the flow path, which are extremely dangerous because of the possibility of resonance with the natural frequencies of various oscillations occurring at hydroelectric power plants [1–3]. The vortex rope significantly narrows the allowable operating range and generates strong oscillations of the entire water column in the penstock, directly affecting the runner [4, 5].

A detailed experimental study of swirling flow behind the turbine runner on full-scale experimental rigs is complicated and expensive. Alternative approaches include the use of scaled-down laboratory models of hydro turbines [6, 7] and simplification of experimental modeling by using air instead of water as a working medium [8–11]. Although these approaches have been developed over many years, modern measurement and design techniques, such as the use of additive 3D printing technologies to create various elements of a turbine model, makes it possible to obtain many more flow regimes and test the geometry of various individual elements (runner, draft tube) in a fairly short time to find the most optimal one.

In our previous work, velocity distributions similar to real turbine velocity distributions were obtained using a system of two vane swirlers in a turbine aerodynamic model [9, 12] and a turbine hydrodynamic model [11, 13]. The present study is an attempt to compare these two cases in terms of

Content from this work may be used under the terms of the Creative Commons Attribution 3.0 licence. Any further distribution of this work must maintain attribution to the author(s) and the title of the work, journal citation and DOI.
Published under licence by IOP Publishing Ltd

available flow characteristics. This is not entirely straightforward since operating parameters such as flow rate and runner rotation speed vary considerably. To solve this problem, we use the integral characteristics of the flow and the dimensionless speed of the runner relative to the optimal operating points of the turbine models.

2. Test rigs

2.1. Turbine aerodynamic model

A detailed description of the aerodynamic setup is presented in [9]. A significant simplification of the experimental modeling is achieved by replacing an aqueous medium with an air medium [14]. The required flow distribution at the diffuser inlet was created using a pair of swirlers: a stationary swirler acting as guide vanes and a rotating swirler—an analogue of a turbine runner. The flow regimes corresponding to the non-optimal (off-design) operating modes of the hydraulic turbine were obtained by adjusting the rotation speed of the movable swirler in the draft tube cone. These flow regimes involve the formation of a pulsating vortex rope, which is of greatest interest. A pair of swirlers was designed for the optimal operating conditions (the best efficiency point, BEP) corresponding to a volumetric flow rate $Q_c = 0.049$ m^3/s and a runner rotation speed $n_c = 2430$ rpm. The Reynolds number ($Re=Q{\cdot}D/A{\cdot}\upsilon$) calculated for Q_c is about $4{\cdot}10^4$. The test section consists of a reduced geometric model of a Francis-99 draft tube with inlet diameter D=100 mm. The flow velocity at local points of the test section was measured with a LAD-06i two-component laser-Doppler anemometer (LDA) with the addition of tracers (small particles of vegetable oil with a diameter of 1 μm) to the air flow. The tracers were formed in a Laskin atomizer. The relative error in measuring the LDA velocity is 0.2%, as follows from the specifications of the instrument. The experimental setup included an automated control system for air flow rate and rotor speed with an error of 1.5% and 0.5%, respectively.

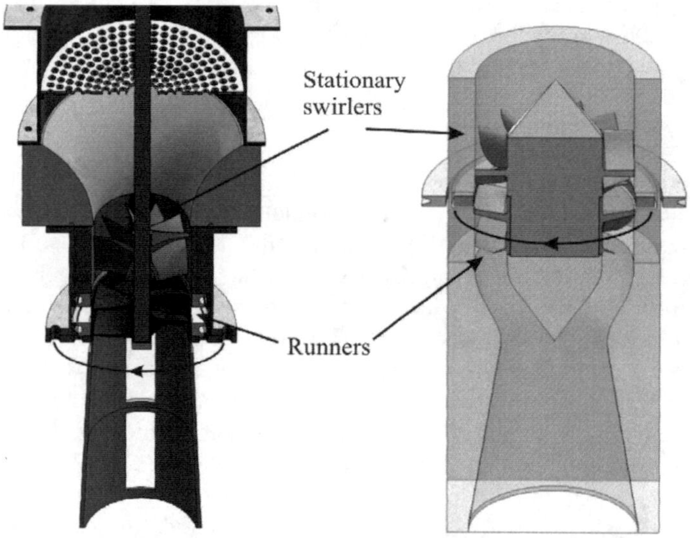

Figure 1. 3D view of the aerodynamic (left) and hydrodynamic (right) turbine models.

2.2. Turbine hydrodynamic model

The turbine hydrodynamic model is also based on the idea of using a swirl generator approach [15] instead of a Francis turbine runner to obtain the same velocity profiles at the draft tube inlet of a Francis turbine operated at partial discharge. This experimental test rig is described in detail in [11, 13]. A 3D view of the test section is shown in Figure 1. The diameter of the draft tube throat is also 100 mm as in the aerodynamic case. The maximum flow rate is up to 0.042 m^3/s (Re = $5{\cdot}10^5$), and the runner speed is up to 1500 rpm. The use of the same measuring equipment as for the aerodynamic setup makes it possible to compare the obtained data with great confidence.

3. Results

In comparing the flow in the investigated turbine models, we will use the main dimensionless quantities characterizing the unsteady turbulent flow:, the Reynolds number (*Re*), the swirl parameter (*S*), the dimensionless flux of moment of momentum (M^*), and the Strouhal number. Despite the fact that the Re number for the air model ($4{\cdot}10^4$) is an order of magnitude lower than that for the water model ($4{\cdot}10^5$), all measurements are in a self-similar region where the dimensionless quantities cease to depend on *Re*.

The measured tangential and axial velocity components at the inlet of the draft tube cone were used to calculate the integral swirl parameter *S* for each flow regime:

$$S = \frac{\int_0^\infty \rho u w r^2\, dr}{R\int_0^R \rho u^2 r\, dr} \qquad (1)$$

This is still one of the main dimensionless quantities characterizing the swirling flow intensity that allows one to roughly compare the operating conditions of swirling devices.

As the runner speed increases, the flow shifts to the walls. At a certain point, the swirling flow is already pressed against the wall; as a consequence, the corresponding swirl parameter stops increasing after reaching a certain saturation level (Figure 2). Thus, the change in the swirl parameter ceases to obey the linear law. The same scenario is observed for the turbine hydrodynamic model. When the dimensionless ratio $N/N_0 = 2.5$ is reached, the integral swirl parameter ceases to increase with increasing dimensionless runner speed and is about of 0.95. For the air model, this value is slightly lower, 0.8, probably due to the difference between the velocity profiles, which, in turn, depend on the shape of the runner blades.

Another parameter by which the flow characteristics can be compared is the Strouhal number ($St = f{\cdot}D{\cdot}A/Q$). Since the dependence of the swirl parameter on the runner speed is nonlinear, the Strouhal number is plotted in Figure 2 as a function of N/N, and not as a function of the swirl parameter$_0$. Figure 2 shows that for both turbine models, the relationship is well described by a linear regression. Moreover, despite the fact that the slope differs starting from $N/N_0 = 3$, the Strouhal numbers take the same values. It is interesting that although for the air model, the vortex precession frequency was measured starting from $N/N_0 = 2$ and for the water model, it was measured starting from $N/N_0 = 1.6$, if we turn again to the swirl parameter, then, due to the difference in the slope of the curves, we obtain the same value $S \approx 0.5$.

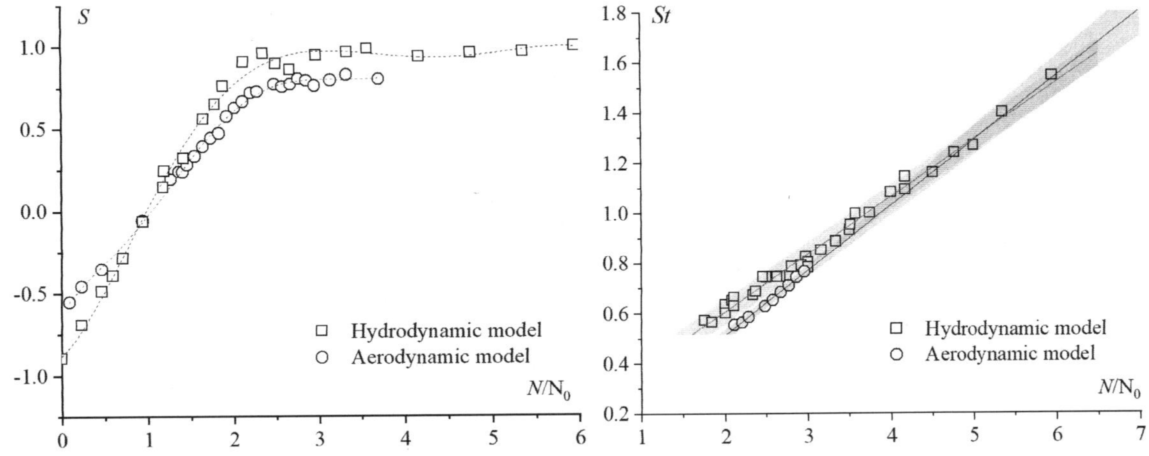

Figure 2. Plots of *S* versus *N/N₀* (left) and *St* versus *N/N₀* (right).

From the point of view of generalization, one can also consider the approach proposed in [16], which allows comparing the dependences of the dimensionless flux of moment of momentum M^* on the dimensionless flow rate Q^*:

$$M^* = \frac{2\pi\omega \int_0^R uwr^2 dr}{(\omega R)^3 \pi R^2} \quad (2)$$

$$Q^* = \frac{Q}{\omega \pi R^3}, \quad (3)$$

where ω is the angular speed of the runner (rad/s), u and w are the axial and tangential velocities, and R is the radius of the runner.

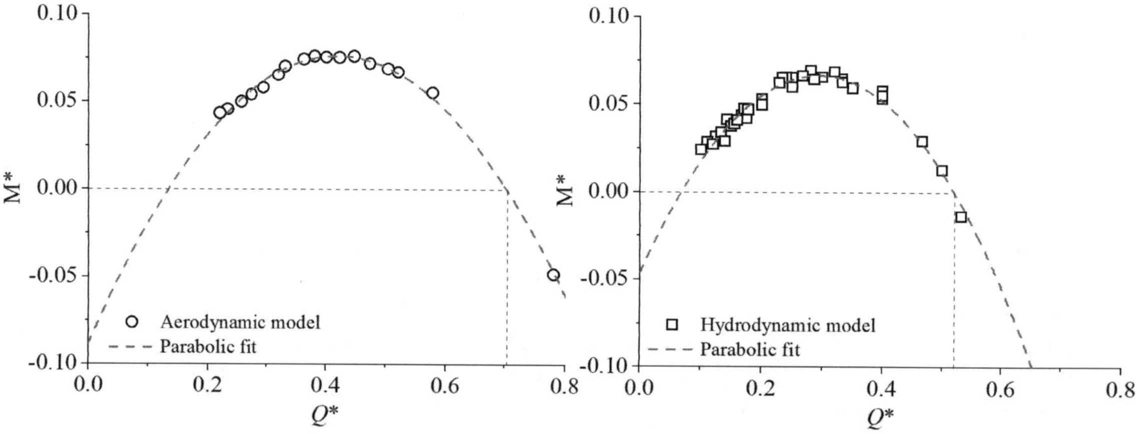

Figure 3. Dimensionless flux of moment of momentum calculated for the aerodynamic (left) and hydrodynamic (right) models.

Figure 3 shows M^* as a function of Q^* for the both cases. One can see that the second-order regression curve fits the experimental data well. The versatility of this approach is that it does not matter under what flow conditions the experiments are carried out. According to [16, 17], the point of intersection of the curve with zero value (M^*=0) corresponds to the optimal operating conditions of the turbine model. In this study, corresponding flow rates Q^* are 0.7 in the case of air and 0.52 in the case of water. A decrease in the dimensionless flow rate relative to the optimal one corresponds to part-load operation. Furthermore, a change in the sign of the dimensionless flux of moment of momentum indicates that the flow is swirled in the opposite direction; i.e. these are forced load conditions. Qualitatively, the curves of the dimensionless flux of moment of momentum look similar, the maximum value of M^* in both cases is approximately equal to 0.075.

4. Conclusion

A comparison was made for hydraulic turbine models operating in air and water. The comparison of the integral characteristics of the flow, such as the impulse flux, and the swirl parameter, as well as the frequency characteristic (the Strouhal number) leads to the conclusion that the replacement of water with air is acceptable for investigating vortex phenomena in swirling flows. The advantages of the air environment include the simplicity and low cost of manufacturing the elements of a model turbine, which makes it possible to sort out a large number of their modifications. It also can be concluded that the effects of cavitation or concentration of dissolved air in the low-pressure region of the PVC observed in the turbine hydrodynamic model play a less significant role in terms of the frequency of pressure pulsations associated with the most powerful vortex mode.

Acknowledgment
The study was supported by a grant from the Russian Science Foundation (Project No. 21-79-10080).

References
[1] Fanelli M 1989 The vortex rope in the draft tube of Francis turbines operating at partial load: a proposal for a mathematical model *J. Hydraul. Res.* **27** 769–807
[2] Susan-Resiga R, Muntean S, Hasmatuchi V, Anton I and Avellan F 2010 Analysis and Prevention of Vortex Breakdown in the Simplified Discharge Cone of a Francis Turbine *J. Fluids Eng.* **132** 051102
[3] Favrel A, Müller A, Landry C, Yamamoto K and Avellan F 2016 LDV survey of cavitation and resonance effect on the precessing vortex rope dynamics in the draft tube of Francis turbines *Exp. Fluids* **57**
[4] Valentín D, Presas A, Egusquiza E, Valero C, Egusquiza M and Bossio M 2017 Power Swing Generated in Francis Turbines by Part Load and Overload Instabilities *Energies* **10** 2124
[5] Müller A, Favrel A, Landry C and Avellan F 2017 Fluid–structure interaction mechanisms leading to dangerous power swings in Francis turbines at full load *J. Fluids Struct.* **69** 56–71
[6] Favrel A, Gomes Pereira Junior J, Landry C, Müller A, Nicolet C and Avellan F 2017 New insight in Francis turbine cavitation vortex rope: role of the runner outlet flow swirl number *J. Hydraul. Res.* 1–13
[7] Goyal R, Cervantes M J and Gandhi B K 2017 Vortex Rope Formation in a High Head Model Francis Turbine *J. Fluids Eng.* **139** 041102
[8] Cassidy, J.J. 1969 Experimental Study and Analysis of Draft-Tube Surging
[9] Litvinov I, Shtork S, Gorelikov E, Mitryakov A and Hanjalic K 2018 Unsteady regimes and pressure pulsations in draft tube of a model hydro turbine in a range of off-design conditions *Exp. Therm. Fluid Sci.* **91** 410–22
[10] Nishi M, Kubota T, Matsunaga S and Senoo Y 1980 Study on swirl flow and surge in an elbow type draft tube *10th IAHR Symposium on Hydraulic Machinery and Cavitation* vol 1 p 557
[11] Skripkin S, Tsoy M, Kuibin P and Shtork S 2019 Swirling flow in a hydraulic turbine discharge cone at different speeds and discharge conditions *Exp. Therm. Fluid Sci.* **100** 349–59
[12] Litvinov I V., Suslov D R, Gorelikov I V., Sadbakov O U and Shtork S I 2018 Swirl number analysis in the air hydro-turbine model *AIP Conference Proceedings* vol 2027 p 040057
[13] Skripkin S, Tsoy M, Shtork S and Hanjalić K 2016 Comparative analysis of twin vortex ropes in laboratory models of two hydro-turbine draft-tubes *J. Hydraul. Res.* **54** 450–60
[14] Nishi M, Yano M, Miyagawa K, Yoshida K, Okamoto M and Liu S 2007 A preliminary study on the swirling flow in a conical diffuser with jet issued at the center of the inlet *Sci. Bull.*

"Politehnica" Univ. Timisoara Trans. Mech. **52(66)** 197–202

[15] Stefan D, Rudolf P, Muntean S and Susan-resiga R F 2013 Structure of Flow Fields Downstream *Eng. Mech.* **20** 339–53

[16] Bosioc A I, Muntean S, Tanasa C, Susan-Resiga R and Vékás L 2014 Unsteady pressure measurements of decelerated swirling flow in a discharge cone at lower runner speeds *IOP Conf. Ser. Earth Environ. Sci.* **22**

[17] Susan-Resiga R F, Muntean S, Avellan F and Anton I 2011 Mathematical modelling of swirling flow in hydraulic turbines for the full operating range *Appl. Math. Model.* **35** 4759–73

Research and results of mathematical modeling of block ion-exchange water treatment plants operating modes and regeneration

A A Chichirov, A A Filimonova, N D Chichirova and O E Babikov

Kazan State Power Engineering University, Krasnoselskaya 51, Kazan, 420066, Russia

aachichirova@mail.ru

Abstract. The problem of water treatment at thermal power plants using ion-exchange technologies is a multi-parameter task. Mathematical modeling is essential for research and optimization of ion exchange technology. The analysis of hydrodynamic processes during the operation of ion-exchange filters was carried out according to the developed mathematical model. Also, a physicochemical analysis of the composition of the water treatment plant solutions under real conditions was carried out. It is shown that in the cationite and anionite filters, the flow movement occurs mainly in a mixed hydrodynamic mode. This mode of regeneration and the filter design do not allow achieving the minimum consumption of the reagent for regeneration, the minimum volume of wastewater and the maximum output of demineralized water. The mixed mode of the anion exchange filter operation allows division of the outgoing solution flow into fractions, which can be successfully used in the TPP water cycle.

1. Introduction

The ion exchange technology is the most widespread at thermal power plants (TPPs) of the Russian Federation due to a good level of ratio between the quality of water treatment and the cost of the process. But at the same time, the ion exchange technology produces the most polluted wastewater at the TPP. Wastewater salinity after ion-exchange plants averages 40-60 g/l. If the ion exchange technology is used at a powerful TPP, then the volume of highly mineralized wastewater, which is 10-15-20% of the input flow, reaches several million tons per year. Such discharges increase the salinity of rivers and lakes and negatively affect the ecology due to deterioration of water quality. The problem of water treatment at thermal power plants using ion-exchange technologies is a multi-parameter task. Under the current operating modes of the ion-exchange chemical-desalting plant at most TPPs, it is impossible to achieve the optimal operating mode. At many TPPs, various technologies have been developed and implemented to reduce the volume of reagents and wastewater. For example, block arrangement of filters of multistage water treatment plants, block-counter-current regeneration, counter-current regeneration of filters, selection and reuse of spent regeneration solutions. Mathematical modeling of ion exchange filters allows you to describe and predict their work with the possibility of correcting the reagent supply of the regeneration mode [1-4].

Earlier [5], a mathematical model of operation and regeneration of an ion exchange filter was proposed and tested on the basis of a one-parameter diffusion model of a flow-through liquid-phase chemical reactor with a continuous loading. It was shown that the regeneration mode implemented in the ion

Content from this work may be used under the terms of the Creative Commons Attribution 3.0 licence. Any further distribution of this work must maintain attribution to the author(s) and the title of the work, journal citation and DOI.
Published under licence by IOP Publishing Ltd

exchange technology and the existing filter designs do not yet allow achieving the theoretically minimum consumption of reagents, the minimum volume of wastewater, and the maximum output of chemically demineralized water.

2. Purpose, materials and methods

The purpose of this article is to study the hydrodynamic and physicochemical processes during the operation of ion-exchange filters and their influence on the composition of the outgoing solutions under real conditions.

We analyze the operation of the cationite H_{II} - H_I - H_{Pr} and anionite An_{II} - An_I semi-chains of the chemical desalting water treatment plant at Nizhnekamsk CHPP-1. According to the existing technology, the chain regeneration operation with a "fresh" acidic solution is carried out along the H_{II} - H_I - H_{Pr} line. The volumetric feed rates of the acidic regeneration solution are 50 m^3/h.

At Nizhnekamsk CHPP-1, the regeneration of the anionite "semi-chain" is carried out with a 2% alkali solution continuously according to the An_{II} - An_I scheme. Filter An_{II} is loaded with a strongly basic anion exchanger, An_I is loaded with a weakly basic one. The diameter of the filters is 3.0 m. The volume of alkali for regeneration is supplied with an excess of approximately two times in relation to the stoichiometric one.

3. Results

The dynamics of the acid output from the H_{II}-filter when supplying "fresh" acid with a volumetric velocity (ϑ) of 50 m3/h is shown in Fig. 1. For comparison, the figure shows the theoretical acid output curves for the ideal mixing mode and the plug flow mode.

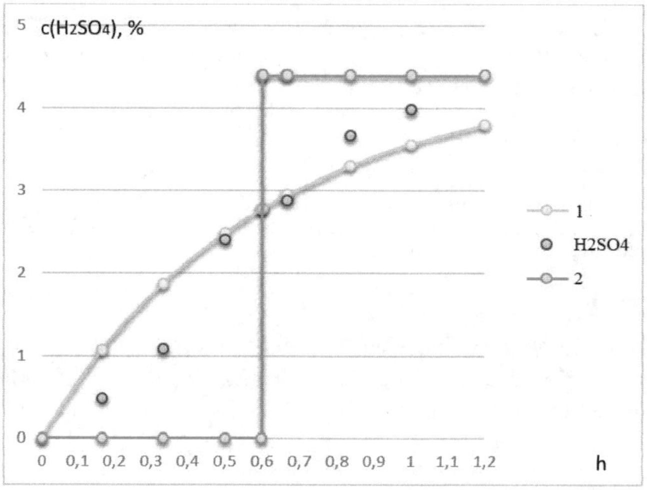

Figure 1. Dynamics of acid output during H_{II}-filter regeneration with 4% H_2SO_4 solution with a volumetric flow rate of 50 m^3/h.
· - experimental values, 1 - theoretical curve for the ideal mixing mode,
2 - theoretical curve for the ideal plug flow mode.

The experimental curve is closer to the theoretical curve of the ideal mixing mode. The theoretical curve characteristics indicate a mixed filter operation, which consists of 75% ideal mixing mode and 25% plug flow mode. Therefore, the regeneration of a single H_{II}-filter will be ineffective in terms of specific characteristics (specific consumption of acid and water), in comparison with the ideal plug flow mode, in which theoretically the minimum consumption of the reagent, the minimum wastewater volume and the maximum product output are possible. The calculated characteristics according to the mathematical model for the regeneration of the cationite "semi-chain" are presented in table 1.

Table 1. Performance characteristics of the H_{II}-filter during acid regeneration with the supply of acid regeneration solution at a volumetric flow rate of 50 m3 / h

Index	Designation and units	Meaning
Characteristic time (average transit time of the filter inlet)	$\bar{\tau}$, h	0.6
Filter transit time by total flow	τ_L, h	0.16
Reduced transit time of the filter by the total flow	θ_L, dm	0.25
Peclet number	Pe, dm	0,33
Counter flow relative to incoming	ϑ_L, m^3/h	$3 \cdot \vartheta_0$

Typical inlet/outlet concentrations of alkali during regeneration are shown in Fig. 2, where the rectangle is the inlet of an alkaline regeneration solution with a constant alkali concentration of 2.4%. At the output from the A_{HII}-filter, the concentration curve is deformed. The blurred edges of the input rectangle, as well as the appearance of alkali at the output of A_{HII}-filter immediately after the start of regeneration, indicate a mixed hydrodynamic mode inside the filters.

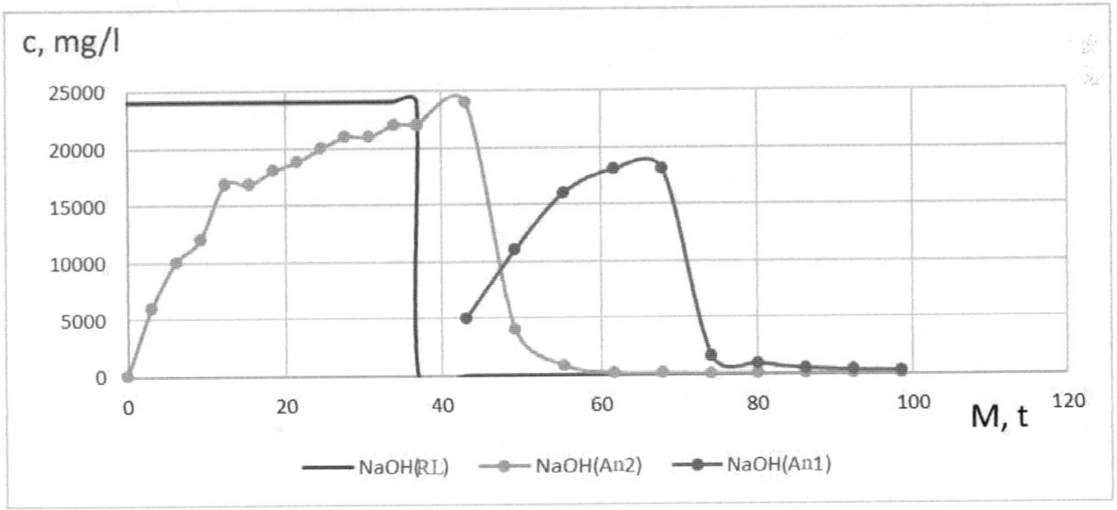

Figure 2. Input/output concentration curves of alkali during regeneration of the "half-chain" of anionite filters No. 8. Consumption of alkaline regeneration solution and partially demineralized water for washing - through A_{HII} with a constant flow rate of 37 t/h. M is the total consumption of solutions through the A_{HII} - A_{HI} chain, tons.

It should be noted that an increase in the flow rate shifts the operation mode towards the plug flow mode, which is consistent with the theory. Comparing the shapes of the output curves of the A_{HII} and A_{HI} filters, we note that in A_{HI} the hydrodynamic mode is closer to the ideal plug flow mode than in A_{HII}.
Regeneration of the A_{HII} filters consumes 6 ± 3% of the initial amount of supplied alkali (Table 2). Up to 50% of the supplied alkali is consumed for the regeneration of A_{HI} filters. The alkali utilization factor for regeneration is slightly more than 50%, the specific alkali consumption is about 80 g/mol-equivalent of anions or 2 mol-equiv / mol-equiv). The losses of alkali used for neutralization is about 50% of the total regeneration consumption. Such results indicate the possibilities of increasing the regeneration process efficiency, both by changing the operating modes and filter design, and by reusing unused alkali.

Table 2. Balance of alkali consumption during regeneration of "semi-chains" of anion exchange filters of water treatment plants at Nizhnekamsk CHPP-1.

The concentration of alkali in the regeneration solution, wt. %	Alkali consumption for regeneration, t/h	Alkali consumption for AH_{II} regeneration,% of the initial	Alkali consumption for AH_{I} regeneration,% of the initial	Unused alkali in % of the initial
2.9	45	11.6	39.8	48.5
2.4	37	5.0	45.4	49.6

The results of mathematical calculations are confirmed by the data of chemical analyzes. The hydrodynamic mode of the ion exchange filters operation is mixed, that is, it occupies an intermediate position between the ideal plug flow and ideal mixing modes. Therefore, the main streams of liquid wastewater from the anionite filter regeneration of a water treatment plant are divided into fractions of approximately equal volume (25 m^3 each) with a characteristic chemical composition (Fig. 3):

- the first fraction is partially demineralized water filling the apparatus after the backwashing;
- the second fraction is the first regeneration product containing mainly sodium salts;
- the third fraction is a contaminated alkaline solution containing organic and silicon compounds desorbed from filters;
- the fourth fraction is an alkaline solution with an impurity concentration of not more than 5% relative to the alkali content;
- the fifth fraction is alkaline wash water with a low content of impurities and alkali.

Analysis of the results using a mathematical model showed that the insufficiently clear separation of the components in the output streams and the diffuse nature of the peaks are associated with a mixed hydrodynamic mode in the filters.

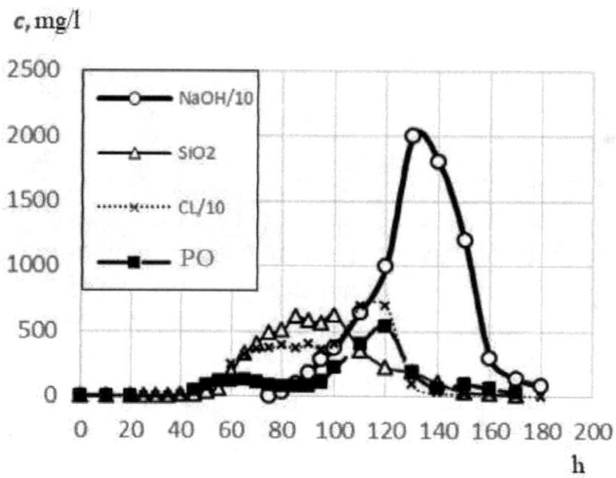

Figure 3. Concentrations of the reagent (NaOH) and desorbed components at the outlet from the "half-chain" (AH_{I})

Taking into account the relatively simple chemical composition and the presence of valuable components in different fractions of liquid regeneration wastes, the possibility of their separation and reuse is indicated.

4. Conclusion

1. The method of mathematical modeling of the flow movement in ion exchange filters during regeneration allows calculating the hydrodynamic operating mode of the water treatment plant, the

composition of the output flows, the half-reaction time of the desorption of substances from the ion exchanger and determining the optimal operating mode that helps to reduce the amount of reagents for regeneration and the amount of wash waters.

2. A mixed hydrodynamic mode of operation is defined in the cation exchange filter, which consists of 75% of the ideal mixing mode and 25% of the ideal plug flow mode. This mode of regeneration and the design of the filter do not allow achieving the minimum consumption of the reagent for regeneration, the minimum volume of wastewater and the maximum output of demineralized water.

3. The mixed mode of the anion exchange filter operation allows division of the outgoing solution flow into fractions, which can be successfully used in the TPP water cycle.

Acknowledgments
The results were obtained with the financial support of the Ministry of Education and Science and the Ministry of Digital Development of Russia in the framework of the agreements terms fulfillment No. 075-15-2021-1087 and No. 075-15-2021-1178 dated September 30, 2021 as part of the strategic academic leadership program implementation "Priority-2030".

References
[1] Safonyk A 2018 Mathematical modelling of regeneration the filtering media bed of granular filters. *Advances in Modelling and Analysis* **2** 72–78
[2] Ravshanov N, Saidov U M 2018 Modelling technological process of ion-exchange filtration of fluids in porous media *J. Phys.: Conf. Ser.* **1015** 1-8
[3] Mukanova B, Glazyrina N 2015 Nonlinear Inverse Problem for an Ion-Exchange Filter Model: Numerical Recovery of Parameters. *Mathematical Problems in Engineering* **8** 11-19
[4] Zhang J, Amini A, O'Neal J A, Boyer T H 2015 Development and validation of a novel modeling framework integrating ion exchange and resin regeneration for water treatment. *Water Research* **84** 255-265
[5] Chichirov A A, Chichirova N D, Filimonova A A, Babikov O E 2020 Mathematical modelling and improvement of ion exchange water treatment energy-saving technologies in power engineering. *Journal of Physics: Conference Series* **15** 1-5

Practical application of natural convective heat exchange of electrical equipment with air

S M Gubanov

National Research Tomsk State University, Tomsk, 634050 Russia

Sgubanov@yandex.ru

Abstract. This paper provides the physical and mathematical model of the air flow in a volume, containing electrical equipment with the heat-generating and heat-absorbing surfaces. The model predicts the temperature fields and air flow velocities across the volume. Using the developed model, we calculate the values of heat fluxes in the vicinity of the thermostated electrical equipment for three different cases: natural convective, forced and mixed modes of the airflow. The possibility of beneficial use of natural convective air flows for the transfer of thermal energy is analyzed. The results are applied in an industrial enterprise. Energy consumption for ventilation is significantly reduced.

1. Introduction

Electrical equipment in most of the cases is cooled by the heat exchange with ambient air [1]. Usually, the approach determining the power of the heat load of the electrical equipment does not consider the influence of the natural convective air flows. The velocity distribution of the airflows near the generating sets and the heat flux values on their surface allow assessing the effect of the natural convection.

Modern methods of the mathematical modeling of air motion provide the values of heat fluxes with high accuracy. The subsequent analysis of the natural convective air motion predicts the effect of convection on the heat transfer intensity. In this paper we theoretically investigate the cooling system of an isotope separation plant as a model case of industrial production with high level of electrical equipment [1]. The aim of the research is the ventilation parameters, which ensures the thermostating conditions of the electrical equipment in the building.

2. Problem formulation

The electrical equipment is placed in the limited area in groups of two in five tiers (Fig. 1). The area is a rectangular parallelepiped, which is usually an element of the periodic equipment arrangement. The height of the region is 7.3 m, the depth is 0.2 m, and the width is 1.5 m. Inside this sample area, we investigate the turbulent natural convective flow of the viscous heat-conducting air. The geometric dimensions of the equipment are stylized with the surface of an average generating set (Fig. 2). The surface of each generating set is divided into sections. The temperatures of the sections are constant, but differ. For the sections of the electric motor winding and the cooling jacket, the temperature is set to the characteristic values.

The heat exchange of air with the walls of electrical equipment is investigated during forced and natural convective air flow in the vicinity of the equipment. The mathematical formulation of the problem takes into account the processes of turbulence using a two-parameter k-ε turbulence model. This turbulence

Content from this work may be used under the terms of the Creative Commons Attribution 3.0 licence. Any further distribution of this work must maintain attribution to the author(s) and the title of the work, journal citation and DOI.
Published under licence by IOP Publishing Ltd

model has limitations when calculating natural convective flows [2]. For this reason, the presented calculations are estimates. The main objective of the presented results is to substantiate a new ventilation scheme for a production facility with electrical equipment.

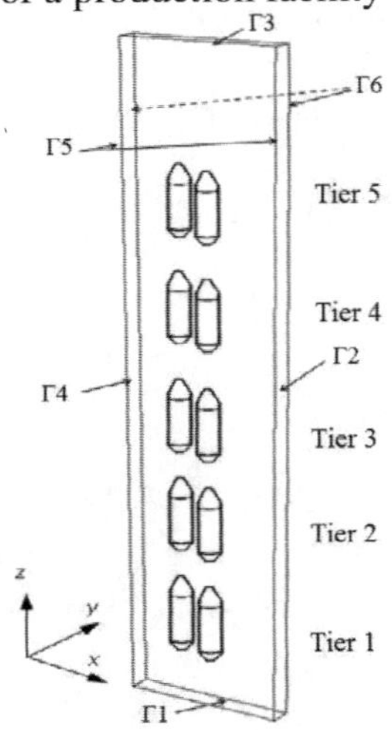

Figure 1. Investigated air flow area

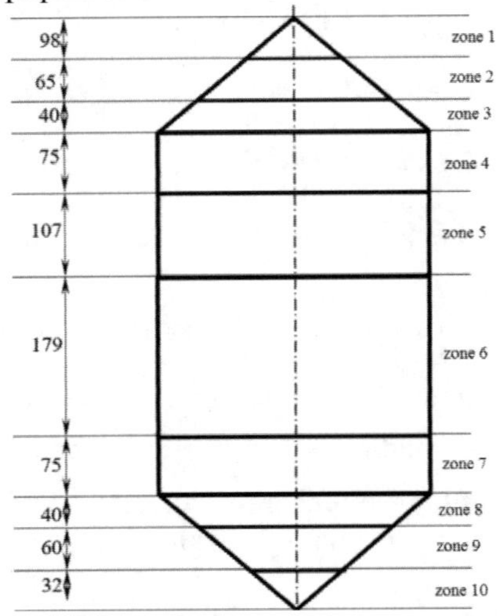

Figure 2. Geometry of the generating set

3. Mathematical model

The stationary natural-convective flow of the viscous heat-conducting air is described by three-dimensional Navier–Stokes equations written for the incompressible gas taking into account turbulence [3]. We use the k-ε turbulence model to describe the turbulent nature of the airflow. The equation system has the following form:

$$\nabla\cdot\left(\rho\vec{V}\right)=0, \quad (1)$$

$$\nabla\cdot\left(\rho\vec{V}\vec{V}\right)+\nabla p=\nabla\cdot\boldsymbol{\tau}_{eff}-\vec{g}\left(\rho-\rho_0\right), \quad (2)$$

$$\nabla\cdot\left(\rho H\vec{V}\right)=\nabla\cdot\left(\lambda_{eff}\nabla T\right)+\nabla\cdot\left(\boldsymbol{\tau}_{eff}\cdot\vec{V}\right)-\vec{g}\left(\rho-\rho_0\right)\vec{V}, \quad (3)$$

$$\nabla\cdot\left(\rho k\vec{V}\right)=\nabla\cdot\left[\left(\mu+\frac{\mu_{turb}}{\sigma_k}\right)\nabla k\right]+G_k+G_b-\rho\varepsilon, \quad (4)$$

$$\nabla\cdot\left(\rho\varepsilon\vec{V}\right)=\nabla\cdot\left[\left(\mu+\frac{\mu_{turb}}{\sigma_\varepsilon}\right)\nabla\varepsilon\right]+C_{1\varepsilon}\frac{\varepsilon}{k}\left(G_k+C_{3\varepsilon}G_b\right)-C_{2\varepsilon}\rho\frac{\varepsilon^2}{k}, \quad (5)$$

$$p=\rho\frac{R_y}{M}T, \quad (6)$$

where: $\vec{V}$ is the gas velocity vector, m/s; H is the total enthalpy, $H=c_pT+\vec{V}^2/2$, J/kg; E is the total energy, $E=c_vT+\vec{V}^2/2$, J/kg; k is the kinetic energy of the turbulent pulsation, m^2/s^2; ε is the dissipation rate of the turbulent kinetic energy, m^2/s^3; $\boldsymbol{\tau}_{eff}$ is the effective stress tensor, $\boldsymbol{\tau}_{eff}=\boldsymbol{\tau}+\boldsymbol{\tau}_{turb}$, Pa; $\boldsymbol{\tau}_{turb}$ is the turbulent stress tensor, $\boldsymbol{\tau}_{turb}=\mu_{turb}\left(\nabla\vec{V}+\nabla\vec{V}^T\right)-\frac{2}{3}\rho k\boldsymbol{U}$, Pa; $\boldsymbol{\tau}$ is the viscous stress tensor,

$\boldsymbol{\tau} = \mu\left(\nabla\vec{V} + \nabla\vec{V}^T\right)$, Pa; μ_{turb} is the turbulent viscosity, $\mu_{turb} = \rho C_\mu \dfrac{k^2}{\varepsilon}$; λ_{eff} is the effective heat conductivity coefficient, $\lambda_{eff} = \lambda + \lambda_{turb}$, W/(m·°K); λ is the thermal conductivity, W/(m·°K); λ_{turb} is the turbulent thermal conductivity, $\lambda_{turb} = \dfrac{c_p \mu_{turb}}{\Pr_{turb}}$, W/(m·°K); G_k is the turbulent kinetic energy generation due to the shearing flow (velocity gradient effect), $G_k = 2\mu_{turb} \boldsymbol{S} : \boldsymbol{S}$, Pa/s; $\boldsymbol{S}$ is the strain tensor, $S = \dfrac{1}{2}\left(\nabla\vec{V} + \nabla\vec{V}^T\right)$; G_b is the turbulent kinetic energy generation due to the natural convection (Archimedes force), $G_b = \vec{g}\dfrac{\mu_{turb}}{\Pr_{turb}}\dfrac{\nabla\rho}{\rho}$; M is the air molecular weight, kg/mol.

The following boundary conditions are set for the equation system (1)–(6): no-slip conditions $\vec{V}(x, y, z, t) = 0$ and conditions for the constant temperature $T = T_{i,j}$ on the walls of the equipment, where *i* corresponds to the number of the section with a constant surface temperature of the generating set, and *j* corresponds to the number of the tier. The turbulent parameters of the flow on the wall of the equipment and the near-wall region are determined as follows: $\dfrac{\partial k}{\partial n} = 0$, $\varepsilon_P = \dfrac{C_\mu^{3/4} k_P^{3/2}}{\kappa y_P}$.

The expression for the rate of the turbulent energy dissipation in a cell near the wall of the equipment is obtained from the expressions for the shear stress on the wall $\tau_w = \mu_{turb}\dfrac{\partial V}{\partial n}$, the turbulent viscosity $\mu_{turb} = \rho C_\mu \dfrac{k^2}{\varepsilon}$ and the logarithmic law of the velocity distribution in the turbulent boundary layer $U^* = \dfrac{1}{\kappa}\ln\left(E \cdot \tilde{y}^*\right)$, where $U^* = \dfrac{C_\mu^{1/4} k_P^{1/2} U_P}{\tau_w / \rho}$, $y^* = \dfrac{\rho C_\mu^{1/4} k_P^{1/2} y_P}{\mu}$, $\tilde{y}^* = \max\left(y^*, y^*_{limit}\right)$; $y^*_{limit} = 11.225$. Here k_p = 0.4187 is the Karman's constant; E = 9.793 is the empirical constant. The value of y^* is determined from the equation:

$$\frac{(T_W - T_P)\rho c_p C_\mu^{1/4} k_P^{1/2}}{q} = \begin{cases} \Pr \cdot y^*, & y^* < y_T^* \\ \Pr_{turb}\left[\dfrac{1}{\kappa}\ln\left(E \cdot y^*\right) + P\right], & y^* > y_T^* \end{cases},$$

where $P = 9.24\left[\left(\dfrac{\Pr}{\Pr_{turb}}\right)^{3/4} - 1\right]\left[1 + 0.28 \cdot \exp\left(-0.007\dfrac{\Pr}{\Pr_{turb}}\right)\right]$.

The value of y_T^* is found by solving the equation: $\Pr \cdot y^* = \Pr_{turb}\left[\dfrac{1}{\kappa}\ln\left(E \cdot y^*\right) + P\right]$.

Symmetry conditions are set on the outer boundaries of the computational domain, being the planes Γ2, Γ4, Γ5, Γ6. The condition $V_n = 0$ is set for the velocity component normal to the boundary, for all other variables – $\dfrac{\partial \varphi}{\partial n} = 0$. Γ1 and Γ3 are the walls. In the case of the forced ventilation of the area, it is assumed that through the border Γ1 air enters the room and leaves through the border Γ3. In this case, the boundary conditions are: $V_z = V_0$; $T = T_0$. The turbulent flow parameters are determined by the intensity of the turbulent pulsations and hydraulic diameter D_{hyd}: $k=\frac{3}{2}(I\cdot V_0)^2$; $\varepsilon=C_\mu^{3/4}\frac{k^{3/2}}{\ell}$; $\ell=0.07\cdot D_{hyd}$. The pressure at the Γ3 boundary is equal to the atmospheric one.

4. Solution method and calculation results

The following thermophysical characteristics of the air are used in the calculations [4]: c_p = 1006 J/(kg·K); λ = 0.0242 W/(m·K); g = 9.81 m/s^2; $\mu = 1.8\cdot10^{-5}$ Pa·s; M= $29\cdot10^{-3}$ kg/mol; Pr_{turb} = 0.85; $C_{1\varepsilon}$ = 1.44; $C_{2\varepsilon}$ = 1.92; C_μ = 0.09; σ_k = 1.0; σ_ε = 1.3.

To solve the equation system (1) - (6), we use the simulation software "Ansys Fluent". The system is solved by the Patankar method using the SIMPLE algorithm. We have carried out the study on the grid convergence. The calculation accuracy is controlled by the balance of the heat and mass fluxes. The computational domain has 960128 cells. The difference grid step size changes from $h_x=h_y=h_z=3\cdot10^{-2}$ м to $h_x=h_y=3\cdot10^{-5}$ м near the boundaries of the heat exchange surface. The chosen difference grid provides the fulfillment of the mass conservation law with the accuracy of 99.7% and the calculation accuracy of the heat fluxes on the surface of the electrical equipment is 96.6%.

Fig. 3 shows the steady-state temperature fields in the vicinity of the electrical equipment. There are two variants of the calculations in the figure: a - natural convective airflow, b - forced air motion. There are two limiting variants of the airflow and heat transfer, presented in Fig. 3.

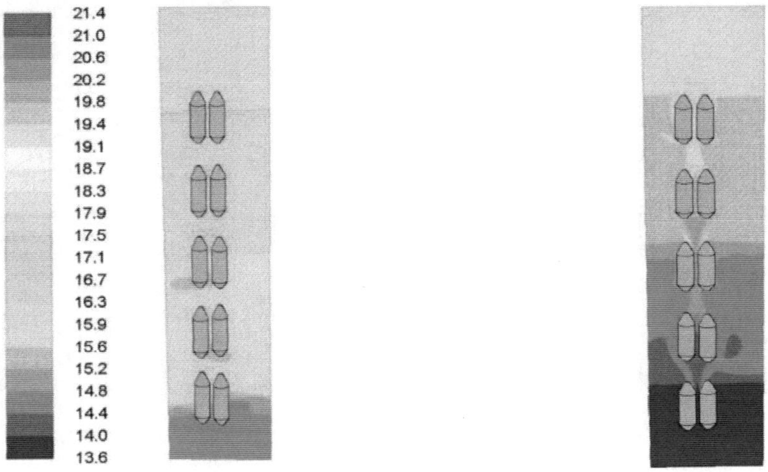

a) natural convective air motion *b) forced air motion*

Figure 3. Temperature field in the vertical middle section of the computational domain

According to the obtained results, the cooling air is stratified in height: colder air is at the bottom and warmer air is at the top (Fig. 3). The greatest temperature gradient in the vertical direction in case of the natural convection is observed in the areas below and above each generating sets (Fig. 3 a). There is a slight change in temperature along the height near the generating sets (Fig. 3a). The air temperature at the top is 290.7 K (Fig. 3a). When the temperature of all the walls of the equipment rises by 2 degrees, the air temperature at the top is 293.47 K. Fig. 3a shows that the upper level of the generating sets is in conditions of the increased air temperature. Overheating of the generating sets is possible in this area. To prevent this effect, it is possible to use intermittent ventilation. When cold air is supplied through the lower boundary of the region, all tiers of the electrical equipment are cooled (Fig. 3b). The air heats up by about 2°C. Both modes satisfy the thermostating conditions of the generating sets. The amount of energy for forced air motion is known from the cost of the ventilation and air conditioning. The question arises: are natural convective airflows capable of removing heat energy and what kind of heat fluxes arise in this case?

Natural convective air motion in the vicinity of the equipment has a complex vortex character. There is a qualitative similarity of the air motion from one level of the electrical equipment to another. The air in contact with the cold areas of the generating set surface descends, whilst the air in contact with warm areas rises. The heated air rises and contacts the cold sections of the next tier of the generating sets, gradually cools down, changes the direction of motion, and starts descending. Natural convective vortices occur between the tiers of the generating sets. The turbulent vortical motion is involved in thermostating of the generating sets. The ascending airflows take part of the generated heat from the surfaces of the generating sets with high temperature and transfer it to the generating sets located above, which have areas with low temperatures at the bottom. The calculated heat fluxes on the surfaces of the electrical equipment are presented in Table 1 and illustrated in Fig.4. Reliable calculation accuracy is about 0.01 W.

The diagrams (Fig. 4) of the heat fluxes distribution on the surfaces of the electrical equipment show the effect of the temperature differences and stratification on the direction and values of the heat fluxes.

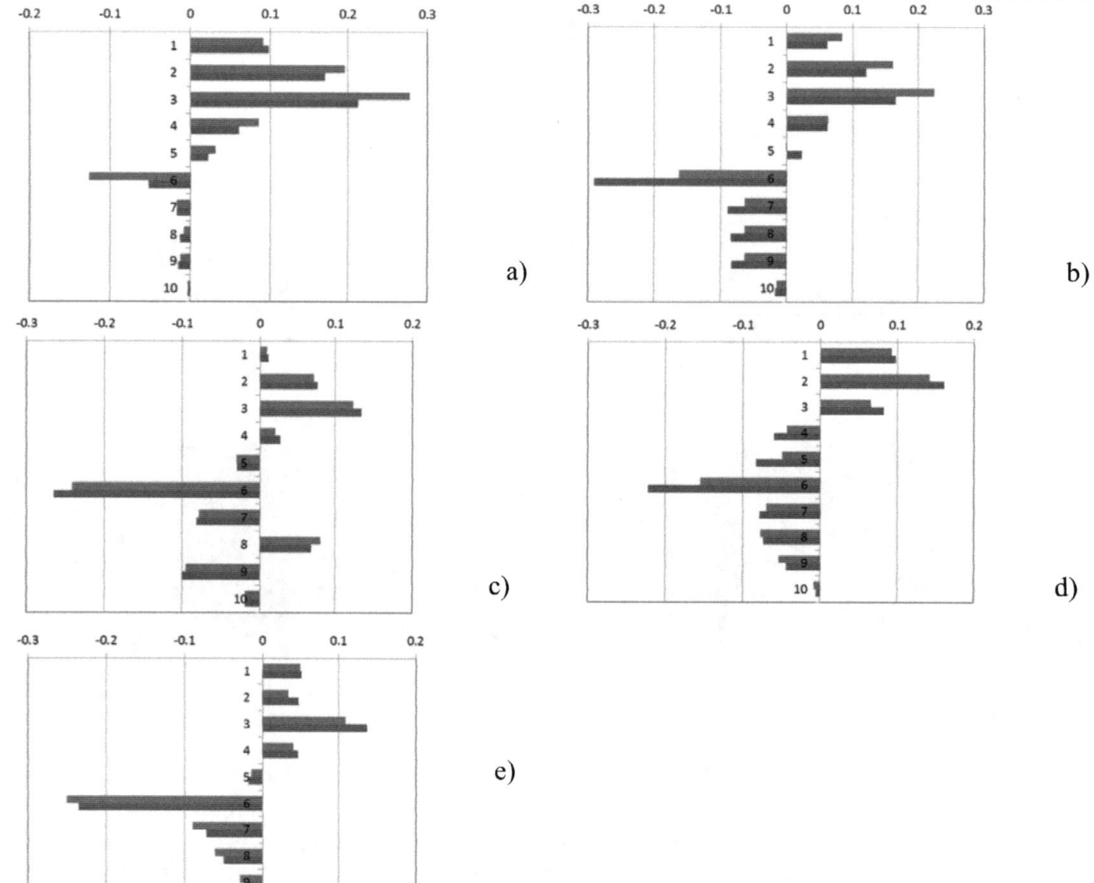

a) – tire 1; b) – tire 2; c) – tire 3; d) – tire 4; e) – tire 5

Figure 4. Heat flux distribution on the surfaces of the generating sets, W

The dark color corresponds to the right row of the electrical equipment, the lighter one - to the left. The positive values correspond to the heat flux into the air, and negative ones – to that from the air. As can be seen from the diagrams, on the lower (first) tier (Fig. 4a), the upper part emits the heat to the surrounding air, whilst the heat flux to its lower part is insignificant. For the generating sets located above, the intensity of the heat flux into the lower part of the equipment increases (Fig. 4 c-e). The air velocity reaches the highest values in the area of 2 and 3 tiers (Fig. 4 b, c). The greatest differences in the direction of the heat fluxes are observed here. Near the fifth tier, this difference decreases.

The calculation results of the total positive and negative heat flux to the generating sets are presented in Table 1. The generating sets of the first tier and the third tier give off heat to the ambient air. Other units receive heat from the air. Heat fluxes for two adjacent pieces of the equipment (left and right rows) are different due to the hydrodynamics of the air flow.

Table 1. Total flux absorbed by the surface of the equipment, W.

Row	Tier 1	Tier 2	Tier 3	Tier 4	Tier 5
Left	0.516906	–0.21986	0.161542	–0.15862	–0.1561
Right	0.467085	–0.11478	–0.13286	–0.17814	–0.22543

The total heat flux from all electrical units into the air is 1.1455 W, and the total heat flux to all electrical units is 1.1858 W. The balance of the fluxes (the sum of the positive and negative fluxes) is -0.0403 W. This value is less than 4% of the total heat flux coming to the electric units from the air. In the absence of the forced air motion, the resulting natural convective movement provides almost zero balance of the heat fluxes: the heat removed into the air from the warm sections of the electrical units equals the heat entering the cold areas of the electrical units. In fact, the air transfers the heat between the surface areas of the generating sets with different temperatures. This mainly occurs between the surfaces of the units located on different tiers. This phenomenon has a significant effect on the thermostating of the electrical units in the absence of the forced air motion.

The numerical modelling of the natural convective air motion around the electrical equipment shows the possibility of thermostating objects without forced convection. It is possible to achieve a zero total balance of the heat fluxes by creating the conditions for the occurrence of natural convection, when the generating sets are located one above the other.

5. Conclusion

Analysis of the results obtained on the natural convective air motion providing the heat exchange between surfaces of the electrical equipment and air shows the possibility to reduce the energy consumption for thermostating objects. It is possible to achieve a complete closure of the heat fluxes within the system by creating the conditions of intensive natural air motion. The values of the air exchange leading to the disappearance of the natural convective air motion have been determined numerically. The developed mathematical model allows investigating the processes of the heat transfer that occurs in real conditions and carrying out the preliminary estimates for the subsequent practical use of the natural convective air flow for thermostating electrical equipment. The method to determine the dependences of the temperature field formation under the natural convective, forced and mixed modes of air motion in the vicinity of the thermostatically controlled electrical equipment has been successfully tested on the operating industrial enterprise - Siberian Chemical Combine.

References:

[1] Gubanov S M, Krainov D A 2012 Proceedings of Tomsk State University 282 (Tomsk: TSU) p 82-86.
[2] Hanjalic K 2002 *Ann. Rev. Fluid Mech* **34** 321-347
[3] Loytsyansky L G 1987 *Mechanics of liquid and gas* (Moscow: Science)
[4] *Handbook on Heat Exchangers* 1987 (Energoatomizdat, Moscow)

Vortex heat transfer enhancement by energy-efficient structured plates with zigzag grooves for micro- and macro-scale energy and electronic devices

S A Isaev[1,2], A I Leontiev[3], D V Nikushchenko[2], D Kong[4], K M Chung[5] and A G Sudakov[1]

[1] St. Petersburg State University of Civil Aviation, St. Petersburg, Russia
[2] St. Petersburg State Marine Technical University, St. Petersburg, Russia
[3] Baumann Moscow State Technical University, Russia
[4] Northwestern Polytechnical University, Xi'an, Shaanxi, China
[5] National Cheng Kung University, Tainan, Taiwan

isaev3612@yandex.ru

Abstract. An energy-efficient flat surface is formed when applying single-row cross-flow zigzag grooves (VVVVVV) in a dense arrangement. Convective heat transfer is considered in turbulent flow around a longitudinal fragment of a plate with a length of 40 and a width of 4 with a package of 14 singlerow inclined backslash(\)-shaped grooves with symmetry conditions at the lateral boundaries. The width of the grooves is 1, the depth is 0.25, the edge rounding radius is 0.2, the angles of inclination are 30°, 45°, 50° and 60°, the pitch is 2.4, the Reynolds number is 104, and the thickness of the incoming boundary layer is 0.175. The phenomenon of anomalous enhancement of the separation flow and heat transfer in zigzag grooves and acceleration of the wall flow discovered in inclined oval-trench dimples has been confirmed.

1. Introduction

Energy efficient surfaces are generally structured. For example, periodic protrusions have been proposed for enhancing turbulent heat transfer in channels and pipes [1]. In this paper, emphasis is placed on structured plates with discrete roughness. The main feature that distinguishes these plates from channels with a constant flow rate is the formation of uniform flow in the space near the plate with a boundary layer growing along the plate. To reduce hydraulic losses (flow resistance of the coolant), reliefs with ordered dimples, mostly spherical, are considered (as, e.g., in [2, 3]). Here, the plate surface is structured with grooves. Analogs are coplanar channels [4]—a well-known method of heat transfer enhancement. However, in contrast to them, the concept of anomalous enhancement of heat transfer in inclined oval-trench dimples (OTD) [5–9] is used in narrow channels and on the ribs of air condensers. The occurrence of a significant static pressure drop between the closely spaced zones of stagnation (positive pressure) of the flow entering the OTD on the windward side and rarefaction (negative pressure) at the site of generation of a tornado-like vortex on the spherical part is due to the ultra-high velocity of the separated and recirculation flow in the OTD. With a dense arrangement of inclined OTD, the maximum velocity of the recirculation flow in the turbulent mode is close to the mass-average velocity, and the velocity of the transverse swirled flow exceeds the maximum velocity in a plane-parallel channel. Acceleration of the turbulent flow is observed in a dimpled narrow channel, and the maximum velocity can be one and a half times higher than the maximum velocity in a plane-

Content from this work may be used under the terms of the Creative Commons Attribution 3.0 licence. Any further distribution of this work must maintain attribution to the author(s) and the title of the work, journal citation and DOI.
Published under licence by IOP Publishing Ltd

parallel channel. In this work, a multi-row relief is formed on the plate by applying transverse zigzag VVVVVV grooves in a dense arrangement.

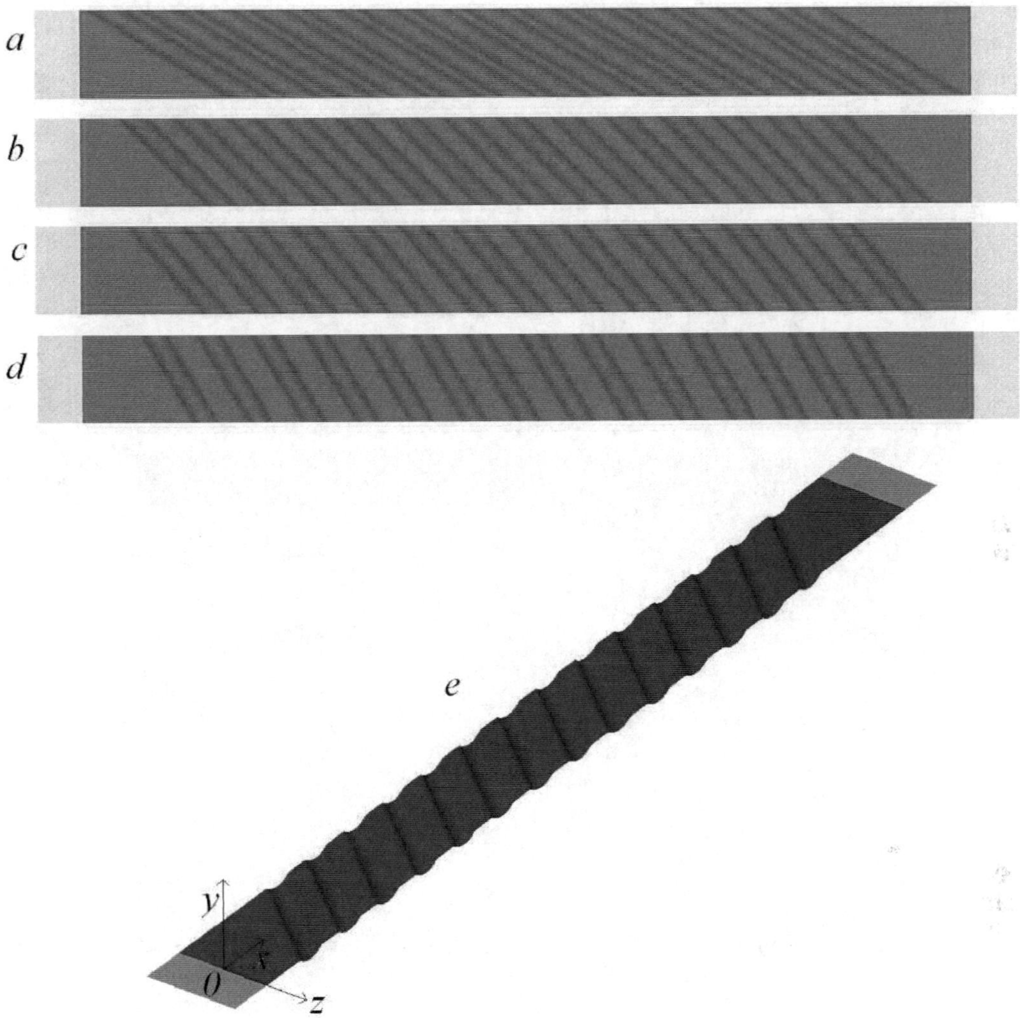

Figure 1. Fragments of a plate with a width of 4 and a length of 40 with backslash(\)-shaped grooves with angles of inclination θ=32° (*a*), 45° (*b*), 50° (*c*), 60° (*d*) (top view) and θ=45° (*e*) (in axonometry, flow from left to right)

2. Statement of the problem and method of solution

The problem of enhancement of convective heat transfer in turbulent air flow around a longitudinal fragment of a plate with a length of 40 and a width of 4 with a dense package of single- row inclined backslash(\)-shaped grooves is posed and solved with symmetry conditions at the lateral boundaries (FIGURE 1). We consider a package of 14 grooves with a width of 1, a depth of 0.25, an edge rounding radius of 0.2, inclination angles of 30°, 45°, 50°, and 60°, and a pitch of 2.4 between the centers of the grooves in the median section of the structured section –. The pre-connected and end sections of the flat plate have a length of 2. The thickness of the boundary layer of the low-velocity air flow at the entrance to the computational domain is set equal to 0.175. The Reynolds number is assumed to be 10^4. The thermal problem is solved separately from the hydrodynamic one. A constant heat flux condition similar to [9] is set on the plate. The Cartesian coordinate system is in the middle of the "inlet" to the dimpled section.

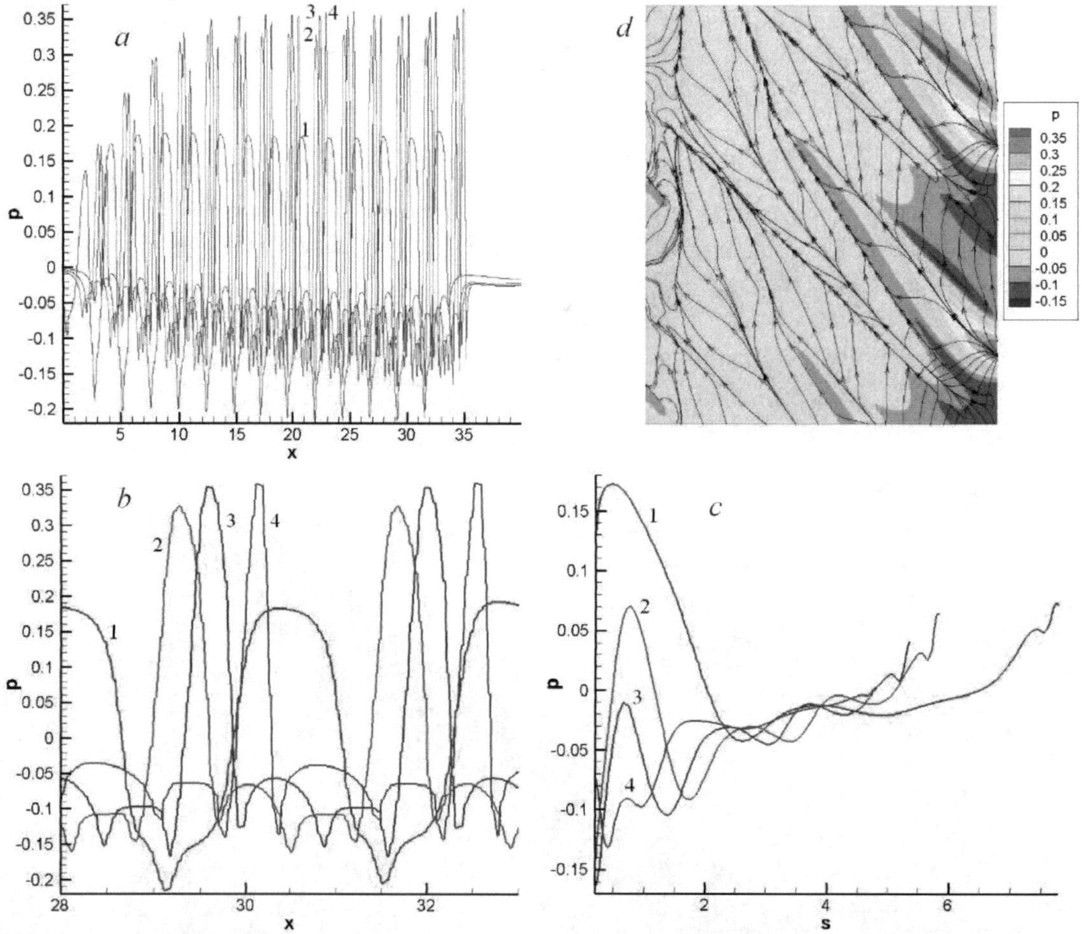

Figure 2. Pressure distributions over the surface of the structured plate in the section z=1.9 (*a*), near the 13*th* groove (*b*), and in the median longitudinal section of the 13*th* groove (c) and the pressure field with a spreading pattern near the 13*th* groove (*d*). 1 – θ=32°; 2 – 45°; 3 – 50°; 4 – 60°

The Cartesian velocity and pressure components are defined in terms of the characteristic velocity. The uniform flow velocity at the inlet to the computational domain is taken as the characteristic velocity. At the lateral boundaries of the computational domain, symmetry conditions are set; i.e., it is assumed that the grooves arranged in one row form a zigzag in the transverse direction. At the outlet of the computational domain, the solution continuation conditions are set. The turbulence characteristics are defined, as in [9], within the framework of the modified Rodi–Leschziner–Isaev SST turbulence model. The factorized finite-volume method implemented in the VP2/3 package is used to solve the original momentum and energy equations. Multi-block overlapping structured grids aligned with a curved wall contain about 6 million cells.

3. Results and discussion

Some results of the numerical study are presented in Figures 2–5 and in Tables 1 and 2. Local characteristics, including extreme ones (Cartesian velocity components u, v, and w, the turbulence energy k, and the vortex viscosity coefficient μ_t), the relative integral force (Cx/Cx_{pl}), and thermal loads (Nu_{mm}/Nu_{mmpl}) at different angles of inclination of the grooves are analyzed.

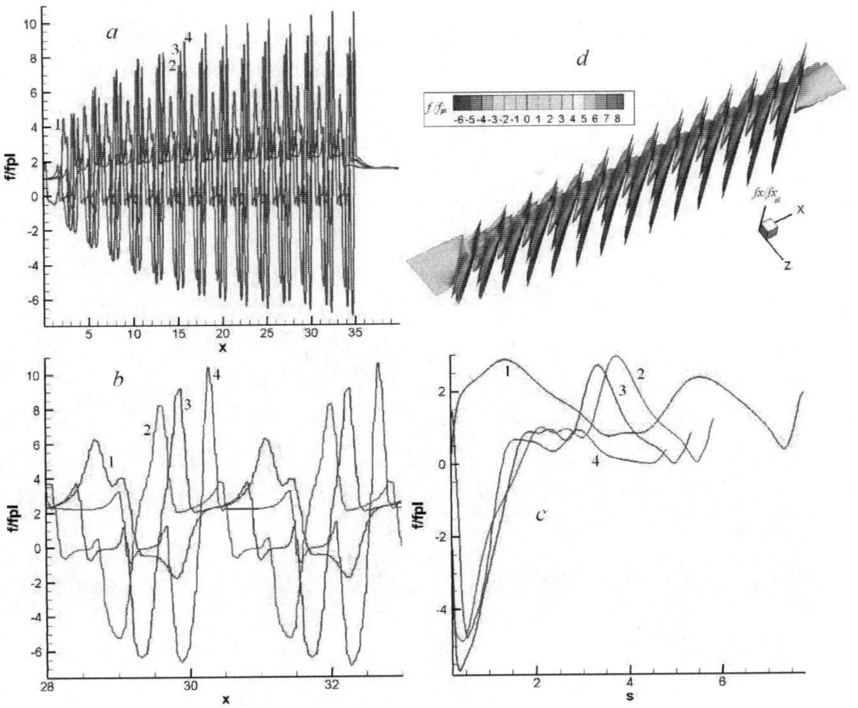

Figure 3. Distributions of the relative friction f/f_{pl} over the surface of the structured plate in the section of z=1.9 (*a*), near the 13*th* groove (*b*), and in the median longitudinal section of the 13*th* groove (*c*) and the field f/f_{pl} (*d*). 1 – θ=32°; 2 – 45°; 3 – 50°; 4 – 60°

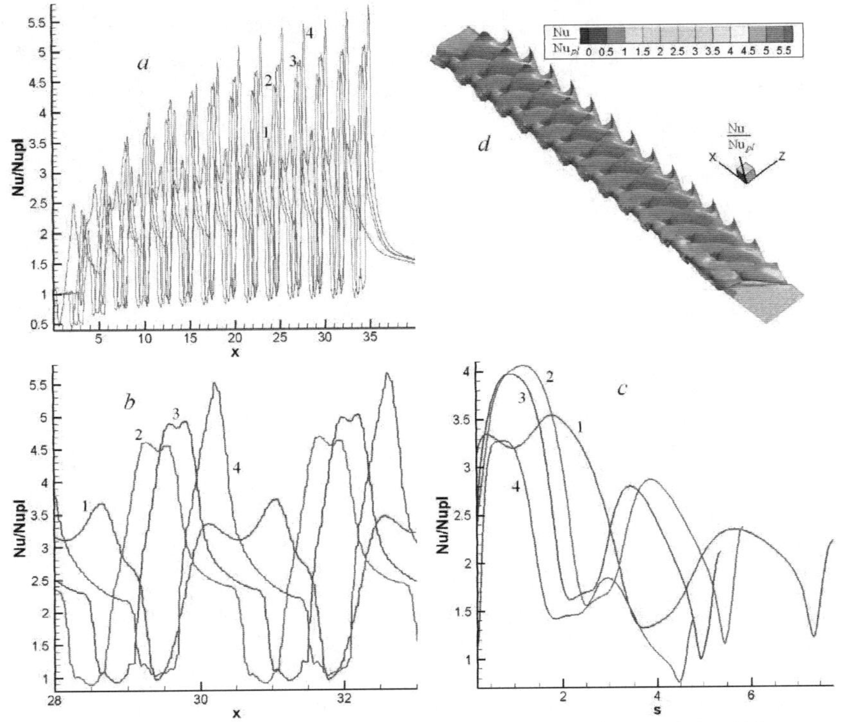

Figure 3. Distributions of the relative heat transfer Nu/Nu_{pl} over the surface of the structured plate in the section z=1.9 (a), near the 13*th* groove (b), and in the median longitudinal section of the 13th groove (c) and the Nu/Nu_{pl} field (*d*). 1 – θ=32°; 2 – 45°; 3 – 50°; 4 – 60°

In the end, the stabilizing turbulent flow and heat transfer in the vicinity of the 13*th* groove are analyzed using the package under consideration. The longitudinal and transverse distributions of the Nusselt number Nu_m averaged over the transverse and longitudinal bands of the structured section of the plate are compared. The thermal-hydraulic efficiency is defined as THE = $(Nu_{mm}/Nu_{mmpl})/(Cx/Cx_{pl})^{1/3}$.

Table 1. Extreme flow characteristics along the structured plate versus the angle of inclination of the grooves

	p_{min}	P_{max}	u_{min}	u_{max}	v_{min}	v_{max}	w_{min}	w_{max}
Plate	-0.012	0.006	0	1.013	-0.001	0.005	-0.00007	0.00007
32	-0.222	0.201	-0.170	1.152	-0.379	0.437	-0.849	0.250
45	-0.346	0.360	-0.486	1.141	-0.396	0.530	-1.015	0.261
50	-0.355	0.392	-0.629	1.148	-0.394	0.557	-1.023	0.252
60	-0.270	0.409	-0.721	1.139	-0.282	0.503	-0.829	0.200

Table 2. Extreme turbulent characteristics, relative coefficients of resistance, and heat transfer of the structured plate and the section of the13*th* groove

10^2k_{max}	$10^2\mu_{tmax}$	Cx/Cx_{pl}	Nu_{mm}/Nu_{mmpl}	THE	Cx_{13}/Cx_{pl13}	Nu_{mm13}/Nu_{mmpl13}	THE_{13}
0.645	0.320	1	1	1	1	1	1
2.348	1.230	3.05	1.87	1.29	3.41	2.29	1.52
3.191	1.776	3.53	1.93	1.27	4.53	2.47	1.50
3.663	1.819	3.45	1.87	1.24	4.55	2.40	1.45
3.471	1.249	2.63	1.58	1.14	3.44	3.44	1.32

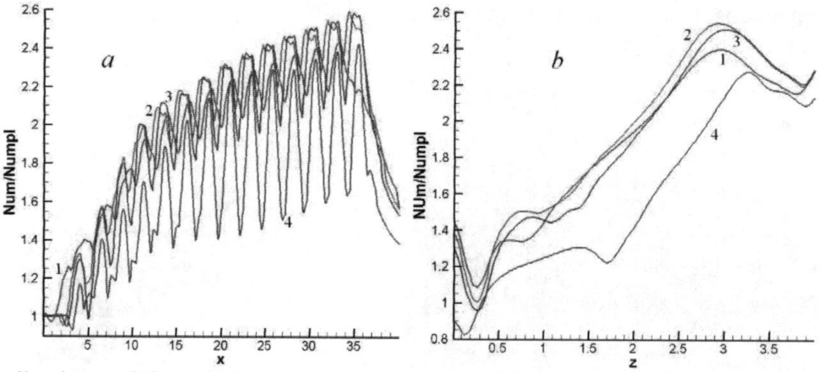

Figure 5. Distributions of the relative Nusselt number Nu/Nu_{mpl} averaged over the transverse (a) and longitudinal (b) bands for different angles of inclination of the grooves. 1 – θ=32°; 2 – 45°; 3 – 50°; 4 – 60°

Interest in the pressure distribution in the longitudinal section near the right boundary of the structured plate section is motivated by the desire to confirm the phenomenon of abnormal enhancement of the separated flow and heat transfer in inclined grooved, previously discovered in inclined oval-trench dimples [5–9]. It is established that this is associated with the occurrence of a pressure drop between the flow stagnation zones on the windward slope and the rarefaction zone at the site of generation of a tornado-like vortex. As can be seen from Figure 2, the pressure drop increases rapidly with distance from the inlet section with an increase in the groove number and stabilizes approximately at the 7*th* groove, reaching 0.5. There is a sharp difference between the pressure distributions at an angle of inclination of 32° and angles of 45° or higher. As θ increases, the rarefaction at the entrance portion of the groove increases. The pressure increases along the groove.

The relative friction, as follows from Figure 3, also sharply increases with an increase in the groove number, but the stabilized mode is reached somewhat later – at the 10–11*th* groove. The value of f/f_{pl}

reaches of 10.5 at the windward ridge of the groove and -7in the region of separated flow. The intensity of the recirculation and secondary flow is maximal at inclination angles of 45–50°, when u_{min} reaches 0.5–0.6 and w_{min} exceeds 1.

The relative Nusselt numbers in Figure 4 also increase as the groove number increases, and at θ=60°, they reach 5.8. It is interesting to note that in the separation zone of the 13*th* groove, Nu/Nu_{pl} is about 4 at θ=45–50°. The relative Nusselt numbers Nu_m/Nu_{mpl} integrated along the transverse bands on the initial hydrodynamic section of the structured plate (FIGURE 5,a) increase in the same way as the considered local characteristics and do not reach stabilization. At θ=45–50°, they reach maximum values of the order of 2.55. The relative Nusselt numbers integrated along the longitudinal bands determine the area of the greatest heat removal on the right half of the plate with the entrance regions of the grooves (FIGURE 5, b). Nu_m/Nu_{mpl} reaches 2.5–2.55 at θ=45–50°. On the left half of the plate, the averaged relative Nusselt numbers are close to 1; i.e., as for packages of inclined oval-trench holes, these zones are not effective.

The thermal efficiency of the structured plate is quite high and reaches 1.93 at θ=45°. However, the relative coefficient of flow resistance turns out to be predominant and equals to 3.53. The obtained data on the resistance are in agreement with experimental results [3]. Nevertheless, the thermal-hydraulic efficiency of the structured plate turns out to be quite high and equal to 1.27. It is interesting to consider the vicinity of the 13th groove in approaching the characteristics of the stabilized flow and heat transfer. The thermal efficiency of the section reaches 2.47, and the relative flow resistance reaches 4.53. The thermal-hydraulic efficiency is higher than in the initial hydrodynamic section and is equal to 1.5.

4. Conclusion

The phenomenon of abnormal enhancement of the separated flow and heat transfer in inclined grooves was confirmed, and the greatest increase in heat transfer of the section with grooves is observed at moderate angles of inclination (45°) in the calculations. The pressure drop, friction, and heat transfer in the grooves and their surroundings increase significantly with distance from the inlet section. The negative friction f/f_{pl} in the separation zones reaches -7, and the positive friction at the rounded edges reaches 10.5. The relative heat transfer Nu_{mm}/Nu_{mpl} from the section with grooves at an angle of inclination of 45° increases to 4.5 at a thermal-hydraulic efficiency of about 1.5. The ratio Cx/Cx_{pl} for the structured plate reaches 3.5 at 45°, which is significantly higher than the twofold increase in Nu_{mm}/Nu_{mmpl}.

Acknowledgments

This research was sponsored by the Russian Science Foundation (Grant No. 19-19-00259, experiment study). The development of the numerical methods was partially supported by the Russian Foundation for Basic Research (Project No. 21-58-52013). The research is partially funded by the Ministry of Science and Higher Education of the Russian Federation as part of World-class Research Center program: Advanced Digital Technologies (contract No. 075-15-2020-903 dated 16.11.2020).

References

[1] Kalinin E et al. 2002 *Efficient surfaces for heat exchangers: fundamentals and design* (New-York, Begell house).
[2] Isaev S, Leont'ev A and Baranov P 2007 *Thermal Engineering* **54** (8) 655
[3] Leontiev A, Kiselev N, Vinogradov Yu et al. 2017 *Int. J. Thermal Sciences* **118** 152
[4] Choi E, Choi Y, Lee W et al. 2013 *Applied Thermal Engineering* **51** 435
[5] Isaev S, Leontiev A, Milman O et al. 2019 *Int. J. Heat and Mass Transfer* **134** 338
[6] Isaev S, Gritckevich M, Leontiev A et al. 2019 *Acta Astronautica* **163** (A) 202
[7] Isaev S, Gritckevich M, Leontiev A et al. 2019 *Int. J. Heat and Mass Transfer* **145** (118737)
[8] Isaev S, Leontiev A, Milman O et al. 2020 *J. Physics: Conf. Series* **1565** 012001
[9] Isaev S, Son E, Leontiev A et al. 2020 *J. Physics: Conf. Series* **1677** 012013

The percolation model of relative conductivities and phase permeabilities

B P Kolesnikov and R R Arushanyan

Kuban State Technological University, Krasnodar, Russia

gradient06@mail.ru, rubenarushanyan@gmail.com

Abstract. The universal mathematical model of relative conductivities of percolation clusters and phase permeabilities of oil-water-saturated rocks is presented. It is obtained on the basis of percolation theory, porous body physics and statistics. The model takes into account the influence of change in pore space surface properties and the nature of fluid flow on the studied characteristics and may be applied for comprehensive analysis and modeling of technological processes of oil production.

The paper is dedicated to the development of the model of relative conductivities of percolation clusters (RCPC) Λ [1, 2], which are used to predict the effective conductivity (heat and electrical conductivity) of heterogeneous environment [3], and relative phase permeabilities (RPP) K [4, 5], used to calculate the dynamics of oil production [6]. The object of the study is oil-water-saturated rocks; their micro- heterogeneous structure is modeled as a set of interpenetrating percolation (infinite) clusters (PC), consolidated from the structural elements (phases, particles, capillaries) of the same-name components and other elements not included in the PC.

Both studied quantities are structural-topological characteristics of the considered environments. Due to the similarity of the transfer processes, they can be presented as a function of the share of the active part of the percolation cluster θ, participating in the transfer process, on the dimensionless critical section α_k of PC, and on the volume concentration of the component ϑ [1, 2]. In the general case for a model of a two-component heterogeneous environment, the function of the RCPC and, hence, the RPP on the specified characteristics may be written as

$$K_{1(2)} \sim \Lambda_{1(2)} = \theta_{1(2)}\alpha_k^{0,5}\vartheta_{1(2)}^{-0,5} = \theta_{1(2)}C_{1(2)}^{0,5}, \quad (1)$$

where $C=\alpha_k/\vartheta$ is the critical adjacency characterizing a degree of phase consolidation.

The quantities included in (1) have all the properties of probability and their values are in the range from 0 to 1, so they can be represented as a power-law relation on the corresponding volumetric content of the component ϑ_i. Critical adjacency can be expressed as the product of the probability of ϑ_i appearance by the probability of preserving it - $C_i=\vartheta_i Z_i$ (Z_i - phase consolidation coefficient, $0 \le Z_i \le 1$) [2]. For a two-component system the condition $Z_1+Z_2=1$ is met, giving the relation that takes into account the interdependence of the critical adjacencies of the corresponding PCs $C_{2(1)}=\vartheta_{2(1)}\left(1-Z_{1(2)}\right)$.

For the case of two-component heterogeneous environment with equal probability of distribution of components' phases (symmetric structure) the model of RCPC [2] is developed on the basis of

Content from this work may be used under the terms of the Creative Commons Attribution 3.0 licence. Any further distribution of this work must maintain attribution to the author(s) and the title of the work, journal citation and DOI.
Published under licence by IOP Publishing Ltd

percolation theory, porous body physics and statistics and the following functions are received: $\Lambda^*_{1(2)} = \vartheta^n_{1(2)}$ - for homogeneous and $\Lambda^*_{1(2)} = \left(\vartheta_{1(2)} - \vartheta_c\right)^t$ – for fractal and quasifractal PC regions, where $t = f\left(\vartheta_{1(2)} - \vartheta_c\right)$ and $n = f\left(\vartheta_{1(2)}\right)$ are critical variable conductivity indices for corresponding PC regions, and ϑ_c is the flow threshold. Taking into account these functions and representing the phase consolidation coefficient as $Z_{1(2)} = Z^*_{1(2)} = \vartheta^m_{1(2)}$ ($0 \le m = f(\vartheta_i) \le 1$, $\vartheta_{1(2)} \le Z^*_{1(2)} \le 1$), it follows from (1) for symmetric structures that

$$\Lambda^*_{1(2)} = \theta_{1(2)}\sqrt{C^*_{1(2)}} = \theta_{1(2)}\sqrt{\vartheta_{1(2)} Z^*_{1(2)}} = \theta_{1(2)}\,\vartheta^{0.5(m+1)}_{1(2)}. \quad (2)$$

The RCPC model for symmetric structures is adapted for environments consisting of unequal components (asymmetric structure), using "oil-water" system in the pore space of oil-water-saturated reservoirs as an example. For such environments, the condition $0 \le Z_{1(2)} \le 1$ that takes into account the surface properties of the pore space must be fulfilled. It is assumed that the critical radius of capillaries $r_k = 2\xi\cos\Theta/\Delta P$ (ξ - coefficient of interphase tension, Θ - wetting angle (contact angle), ΔP - pressure difference in the fluids [4]) determines the critical section - $\alpha_k \sim r_k^2$, and therefore the critical adjacency $C \sim r_k^2/\vartheta$. Then the ratio of the areas of the critical section of the fluid, respectively, at $0^\circ \le \Theta \le 90^\circ$ and $\Theta = 0$ gives the expression

$$\beta = (\cos\Theta)^2, \quad (3)$$

which was used to calculate the PC's phase consolidation coefficient of both asymmetric and symmetric structures in the form - $Z_{1(2)} = \vartheta^{m+\beta/(1-\beta)}_{1(2)}$. It follows from (3) that at $\Theta = 90^\circ$ ($\beta = 0$) the value $Z_{1(2)}$ corresponds to neutral wettability, and hence to the symmetric RCPC model - $\vartheta_{1(2)} \le Z_{1(2)} = Z^*_{1(2)} \le 1$. For hydrophilic and hydrophobic cases (asymmetric system) the wetting fluid angle $0^\circ \le \Theta_{wf} < 90^\circ$ is taken into account ($0 < \beta \le 1$, $0 \le Z_{1(2)} < 1$).

Hydrophilic environment Hydrophobic environment

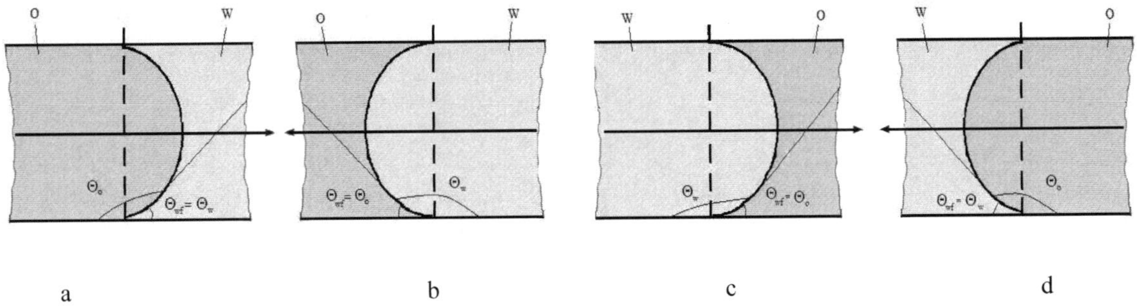

Figure. 1. Profiles of oil and water contact surface in the capillaries of hydrophilic (a, b) and hydrophobic (c, d) environment: a, c – drainage; b, d - impregnation.

Fig. 1 shows capillary profiles in the area of oil and water contact, where the wetting angles used to calculate by formula (3) for drainage and impregnation for hydrophilic and hydrophobic cases are shown.

From comparison of RCPC formulas (1) and (2) it follows that

$$\Lambda_{1(2)} = \Lambda^*_{1(2)}\sqrt{Z_{1(2)}/Z^*_{1(2)}} = \Lambda^*_{1(2)}\,\vartheta_{1(2)}^{0.5\beta/(1-\beta)}. \quad (4)$$

The above dependences allow calculating the RCPC in the following sequence: the wetting fluid angle Θ_{wf} is set; the value of β is defined by formula (3); for the wetting fluid - RCPC Λ_{wf} is calculated by formula (4) using the RCPC Λ^*_{wf} for the symmetric model and $Z_{wf} = Z^*_{wf}\,\vartheta_{wf}^{\beta/(1-\beta)}$ is calculated; for the non-wetting fluid - $Z_{nwf} = 1 - Z_{wf}$ is found and the RCPC Λ_{nwf} is calculated by expression (4).

When applying the considered model to predict RPP, the relation of fluid-conductivity and the critical capillary radius characteristic of the Poiseuille flow ($K \approx r_k^4$) is taken into account [7]. It follows that: $\alpha_k = \alpha_{Kk} \sim r_k^4$ ($C_{K1(2)} = \alpha_{Kk}/\vartheta \sim r_k^4$) and $C_{K1(2)} = C^2_{1(2)}\vartheta_{1(2)}$, which allows finding the correlation between RCPC and RPP according to expression (1)

$$K_{1(2)} = \theta_{1(2)}\sqrt{C_{K1(2)}} = \Lambda_{1(2)}\sqrt{C_{1(2)}\vartheta_{1(2)}}\ . \quad (5)$$

When analyzing the results obtained using the considered model for the process of two-phase (oil - water) filtration in the formation, on the basis of the formula (5), the emphasis is placed on prediction of RPP, for which there is sufficient information on experimental and theoretical data for verification [4, 5, 8]. Hereinafter, RCPC, RPP and other parameters for water are written with the index "w" (Λ_w, K_w, Θ_w, ϑ_w), and for oil - "o" (Λ_o, K_o, Θ_o, ϑ_o).

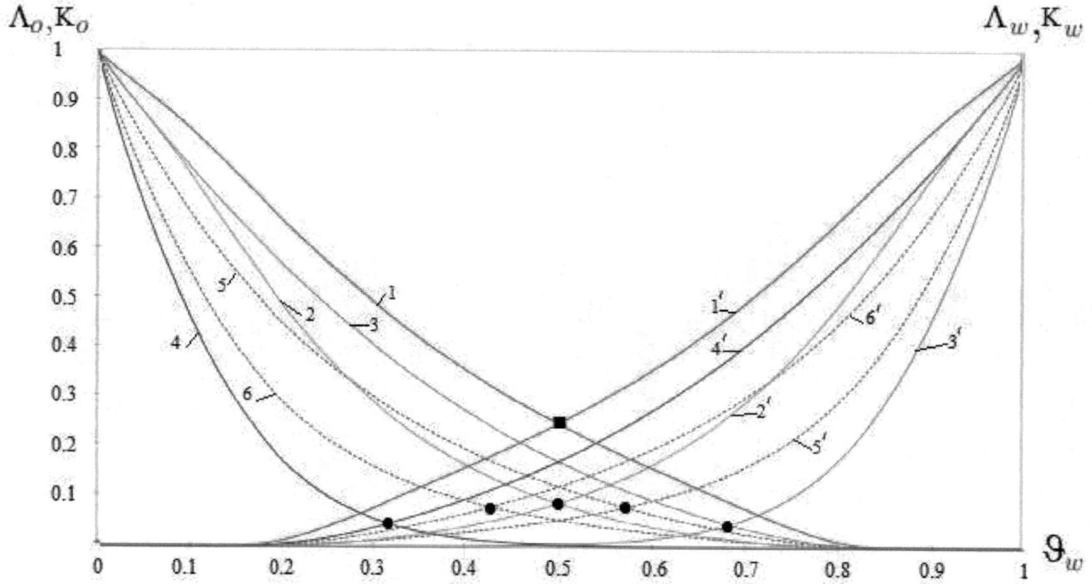

Figure. 2. Functions of RCPC and RPP for "oil-water" environment on water-saturation for strongly hydrophilic and strongly hydrophobic rocks, obtained using percolation model of RCPC and RPP: 1, 1′ – RCPC, the rest – RPP, solid lines – drainage, dashed lines - impregnation; 1 - 6 – oil; 1′ - 6′ – water; 1, 1′, 2, 2′ – $\Theta_w = 90°$; 3, 3′ – $\Theta_w = 25°$; 4, 4′ – $\Theta_w = 155°$; 5, 5′ – $\Theta_w = 25°$ and $\chi = 0.7$; 6, 6′ – $\Theta_w = 155°$ and $\chi = 0.3$;
■ - corresponds to the condition $\Lambda_w = \Lambda_o$, ● - corresponds to the condition $K_w = K_o$.

Fig. 2 shows the calculated functions of RCPC and RPP on water-saturation ϑ_w obtained using the proposed model for strongly hydrophilic ($\Theta_w = 25^\circ$), neutral ($\Theta_w = 90^\circ$) and strongly hydrophobic ($\Theta_w = 155^\circ$) rocks. The obtained functions confirm Craig's rule [9]: by condition $K_w = K_o$ water-saturation for strongly hydrophilic systems is $\vartheta_w > 0.5$, for strongly hydrophobic - $\vartheta_w < 0.5$, for neutral - $\vartheta_w = 0.5$; initial (critical) water-saturation is around 0.16 - 0.2 for strongly hydrophobic systems and above 0.25 for strongly hydrophilic systems. The RCPC curves (1, $1'$) are higher than the RPP curves (2, $2'$) corresponding to the same wetting angle ($\Theta_w = 90^\circ$), which follows from expression (5).

The presented model covers almost all possible variants of water-oil systems arising during drainage and impregnation for hydrophilic and hydrophobic systems, what allows taking into account the hysteresis phenomenon of RCPC and RPP arising at a change of direction of filtration flows [4, 5].

During drainage (displacement of the wetting fluid by the nonwetting fluid), for the hydrophilic case (Fig. 1a) the calculated angle of the wetting fluid (water) is $\Theta_w < 90^\circ$, for the hydrophobic one (Fig. 1b) the calculated angle of the wetting fluid (oil) is $\Theta_o < 180^\circ - \Theta_w$.

In case of impregnation (displacement of nonwetting fluid by wetting fluid), at a decrease of degree of capillary surface hydrophilicity, values of interphase contact degree for oil and water are calculated as at drainage, but with larger angle Θ_w for hydrophilic systems, and with larger angle Θ_o for hydrophobic ones. The calculated angle is $\Theta_o < 180^\circ - \Theta_w$ in case of surface state changing from hydrophilic to hydrophobic (Fig. 1c), and $\Theta_w < 90^\circ$ in case of changing from hydrophobic to hydrophilic (Fig. 1d). The differences in the drainage and impregnation processes disappear when $\Theta_w = \Theta_o = 90^\circ$ (curves 1, $1'$, 2, $2'$ in Fig. 2).

In the process of impregnation the surface properties of capillaries can change, which leads to microheterogeneous wettability. This phenomenon can be taken into account by means of approach [4] consisted in grouping capillaries into two groups: the first group includes capillaries with unchanged properties and has relative fraction χ, and the second group includes capillaries with changed properties and, respectively, has fraction $1-\chi$. Fig. 2 shows RPP functions for impregnation process (5, $5'$, 6, $6'$) calculated considering division of capillaries into two groups with greatly differing surface properties: with wetting angles $\Theta_w = 25^\circ$ and $\Theta_w = 155^\circ$, respectively, with fractions χ and $1-\chi$. It is obvious that the system passes from strongly hydrophobic to strongly hydrophilic state in the limit case when χ changes from zero to one, herewith the value of water-saturation corresponding to $K_w = K_o$ increases from $\vartheta_w \approx 0.3$ to $\vartheta_w \approx 0.7$.

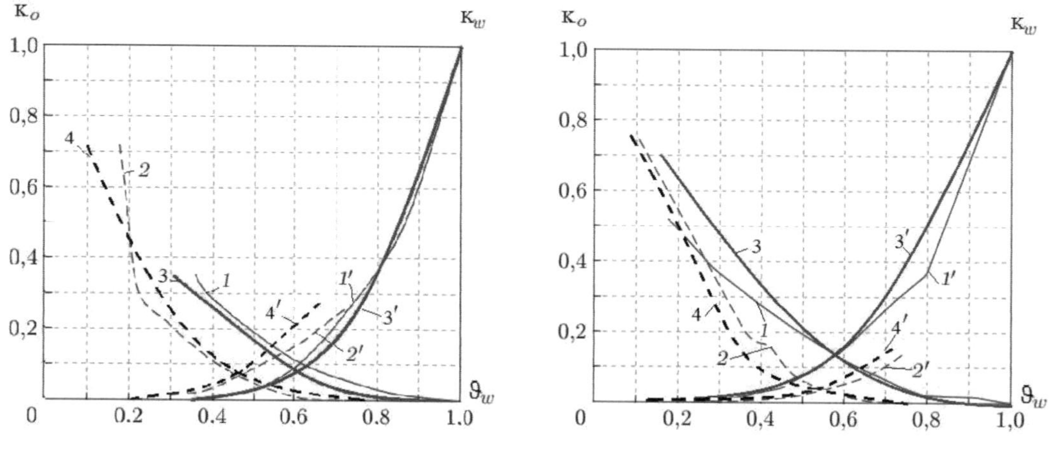

a b

Figure. 3. Functions of RPP and water-saturation obtained with the help of percolation model of RCPC and RPP and experimental data [5] for dolomite (a) and calcite (b) samples: solid lines - drainage, dashed lines - impregnation; 1 - 4 - oil; $1'$ - $4'$ - water; 1, $1'$, 2, $2'$ - experiment [5]; 3, $3'$, 4, $4'$ - calculation by the proposed model: a) drainage - $\Theta_{wf} = \Theta_w = 38^\circ$, impregnation - $\chi = 0.2 \rightarrow \Theta_{wf} = \Theta_w = 38^\circ$, $1-\chi \rightarrow \Theta_{wf} = \Theta_o = 47^\circ$, $\Theta_w = 133^\circ$; b) drainage - $\Theta_{wf} = \Theta_w = 43^\circ$, k=2, impregnation - $\chi = 0.25 \rightarrow \Theta_{wf} = \Theta_w = 43^\circ$, k=2, $1-\chi \rightarrow \Theta_{wf} = \Theta_o = 85^\circ$, $\Theta_w = 95^\circ$, k=0.3.

For complex microheterogeneous wettability in the proposed model it is suggested to form capillaries with equal surface properties by groups in fractions of total number (χ_i - fraction of the i-th group, $\sum_1^n \chi_i = 1$), and to calculate RCPC and RPP of each group using appropriate calculated wetting angle Θ_i. The resultant values of RCPC and RPP are obtained by summing the RCPC and RPP of all groups, taking into account the weight of each: $\Lambda_{w(o)} = \sum_1^n \chi_i \Lambda_{w(o)}(\Theta_i)$, $K_{w(o)} = \sum_1^n \chi_i K_{w(o)}(\Theta_i)$.

In order to more accurately approximate the calculated relations to the experimental ones, there is an extension of the model by introducing a correction that takes into account changes in the rheological properties of oil and water, occurring, for example, during RPP hysteresis. The correction is based on the character of fluid-conductivity dependence on critical capillary radius at flow of fluids with different properties. In view of this, the adjacency used in formula (5) is expressed as $C_{K1(2)} = \vartheta_{1(2)}^{-1} \left[C_{1(2)} \vartheta_{1(2)} \right]^{1+k^{-1}}$, where k is the flow index [5]).

The considered model is verified through fitting its parameters by comparing the obtained calculated values of RPP on water-saturation with the known experimental and calculated functions [4, 5, 8]. As an example, Fig. 3 shows experimental and calculated (by the proposed method) relations of RPP on water-saturation for dolomite and calcite samples, normalized to the absolute permeability of the environment for the same phase [5]. Given the level of reliability of the experimental data, the results of verification can be considered satisfactory.

The presented model shows a generality, as it allows comprehensively analyzing and predicting the RCPC and RPP by selecting optimal values of phase consolidation coefficients, used in calculations, by comparing with experimental data. In consequence of similarity of RCPC and RPP in a number of cases within the presented model, it seems reasonable to use values of RPP obtained by experimental or analytical methods in calculations of effective thermal conductivity of oil-water-saturated rocks.

In contrast to the known percolation models of RPP hysteresis [4, 5], the advantages of the presented model are the simpler mathematical apparatus used without qualitative losses, the complexity of the study of RCPC and RPP, the prospect of taking into account various factors affecting the phase consolidation processes and the possibility of application to study other transfer properties.

References

[1] Kolesnikov B P Influence of fractal substructures of the percolating cluster on transferring processes in macroscopically disordered environments // Journal of Physics: Conference Series Volume 891 (2017) 012355
[2] Kolesnikov B P The unified approach to a definition of effective conductivity index of percolating cluster in macroscopically disordered environments // Journal of physics: conference series : All-Russian scientific conference with international participation "Thermophysics and Power Engineering in Academic Centers" (TPEAC-2019)
[3] Kolesnikov B P Prediction of effective conductivity of multicomponent macroscopically disordered environments by mathematical modeling // Electronic multidisciplinary journal “Scientific works of KubGTU” (2016) № 16 – p. 459-468
[4] Kadet V V, Galechyan A M Percolation model of relative phase permeability hysteresis // Applied Mechanics and Technical Physics (2013) V54 № 3 – p. 95-105
[5] Kadet V V, Galechyan A M Consideration of fluid rheology in the hydrophobization model of relative phase permeability hysteresis hysteresis // Applied Mechanics and Technical Physics (2017) V58 № 6 – p. 58-68
[6] Gudok N S, Bogdanovich N N, Martynov V G Determination of physical properties of oil-bearing rocks // Nedra (2007) – p. 592
[7] Landau L D, Livshic E M Theoretical Physics: // Textbook VI. Hydrodynamics. 3rd ed. Moscow (1986) – p. 736
[8] Moiseev V D Experimental determination of relative phase permeabilities for oil and gas reservoirs in the fields of northern West Siberia // Oil and Gas (2020) – p. 46-53
[9] Graig F F The reservoir engineering aspects of wettability on reservoir rocks and its evaluation // Prod. Monthly (April 1968) V 32 № 4 p/ 2-7

On thermal problems of the solar thermal multigeneration: new nanomaterials and working fluids

A S Dmitriev [1] and A V Klimenko [2]

[1] National Research University «MPEI», Russia, 111250 Moscow, 14 Krasnokazarmennaya Street,
[2] National University of Science and Technology «MISIS», Russia, 4 Leninskiy Prospekt, Moscow

asdmitriev@mail.ru

Abstract: The study considers various thermophysical problems, including the processes of heating new nanomaterials and nanofluids, as applied to solar thermal multigeneration (steam generation and turbine and turbineless conversion into electrical energy), as well as thermoelectric generation (solar thermoelectric generation) using new nanomaterials, including graphene and its composites. The main unsolved problems of thermohydrodynamics and heat transfer in such systems are noted.

1. Introduction

The powerful turbine conversion power plants have known disadvantages, primarily associated with a large number of optical concentrators, the need for precise positioning of concentrated radiation on a container with a working fluid. Failure to meet one of these conditions known to lead to serious accidents, as happened at the power unit mentioned in [1-4].
Along with the purely thermal turbine conversion of solar radiation of the Planck spectrum, other types of conversion have been discussed in recent years, including those associated with multigeneration - the simultaneous use of not only a steam turbine, but also, for example, thermoelectric conversion [1,2].

2. The main thermophysical problems of solar multiconversion

In recent years, interest in solar thermal power engineering based on new materials and technologies has sharply increased [1-6]. The efficiency of various types of generation due to the conversion of solar radiation of the Planck spectrum is largely related to both the use of conversion technologies (steam generation - volumetric and surface with subsequent conversion of steam through machine (turbine) and electric generator conversion), and the thermoelectric generation, i.e., conversion of solar radiation heat with a high degree of its absorption in nanomaterials due to the use of modern and promising materials with high quality factor, as well as with hybrid conversion of solar radiation (part - photoelectric conversion, part - steam generation due to heating by Planck spectrum radiation, followed by thermoelectric or steam generation). Finally, the use of materials with phase transitions (for example, paraffin-graphene composites and their analogs), followed by the use of the heat of phase transitions in energy-efficient converters of low-grade heat or according to a thermo-electrical scheme. However, all the noted transformation methods can be combined and used as multi-

Content from this work may be used under the terms of the Creative Commons Attribution 3.0 licence. Any further distribution of this work must maintain attribution to the author(s) and the title of the work, journal citation and DOI.
Published under licence by IOP Publishing Ltd

generation, the efficiency of which is related to the calculation of the efficiency of each of the channels. Studies have shown that in such a scheme there may be a synergistic effect, in which the total efficiency is not additive in terms of contributions from certain types of solar radiation conversion [7] (Figure 1).

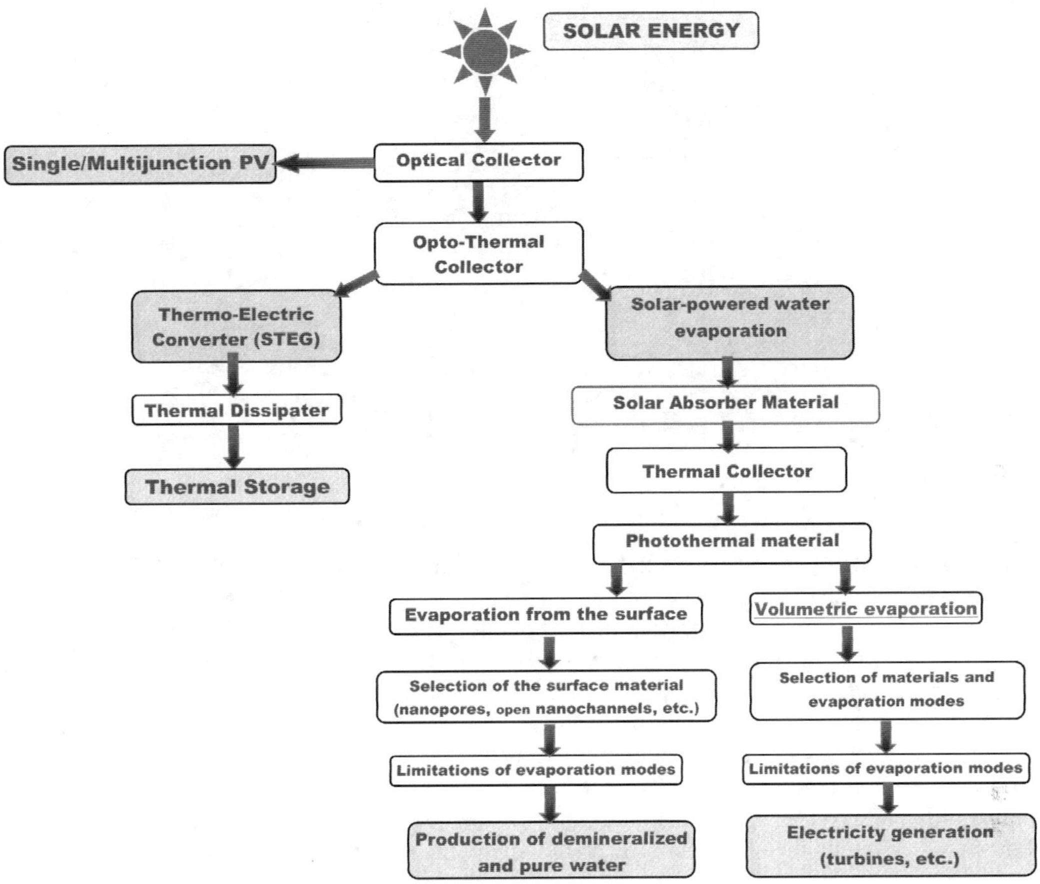

Figure 1. Solar thermal multigeneration

2.1. Solar steam generation

In such problem statement, thermophysical problems play a decisive role, since the indicated efficiency largely depends on the mechanisms of absorption, distribution and transfer of heat [1]. At present, not enough attention is paid to works devoted to both the calculation of individual processes and model experiments on solar multigeneration. At the same time, it is well known that the first stage of considering such problems is directly related to the choice of materials that should have a number of important parameters. These parameters primarily include: high absorption coefficient of solar radiation, high heat capacity and thermal conductivity, low thermal boundary resistance, etc. Among such materials, in recent years, close attention has been paid to nanomaterials, in particular two-dimensional ones, including graphene and its nanocomposites, as well as nanofluids.

Solar thermal multigeneration can include several channels for using the solar radiation spectrum: from the optical absorption of photons with energy greater than the band gap – PV cells, to the almost complete absorption of the entire Planck radiation spectrum – thermoelectric (STEG) and machine (turbine or turbineless) conversion (Figure 2).

Numerous works on the selection of new nanomaterials and nanocomponents for intensifying the evaporation of water and generating steam with high efficiency have appeared recently [1-4]. In particular, research is being actively carried out on graphene nanoporous structures, which show an extremely high degree of absorption of solar radiation, rapid heat transfer into the material, and a high evaporation rate (Figure 3).

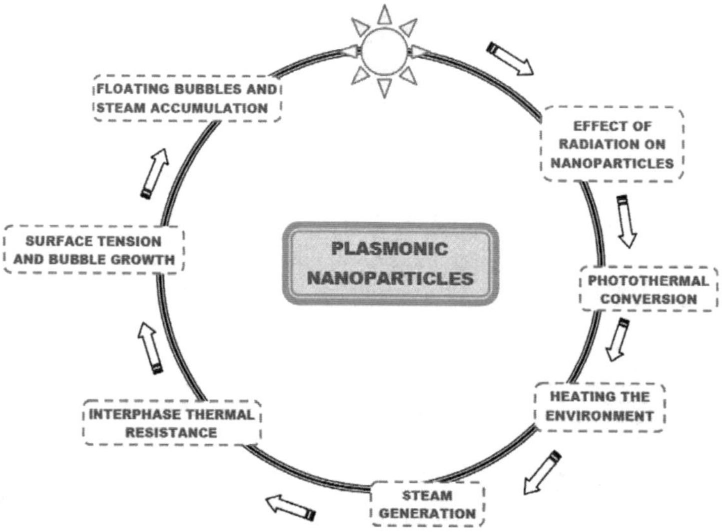

Figure 2. Steam generation processes using nanoplasmonics and nanophotonics

This work formulates the main criteria for solar multigeneration, and some problems related to the mechanisms of radiation absorption in nanomaterials, heat transfer mechanisms and boundary thermal resistance, which significantly expand the class of heat transfer mechanisms in nanostructures [1]. In addition, the main advantages of using new thermoelectric materials in multigeneration at various temperature levels are indicated. For the first time, calculations of the conversion efficiency of solar radiation to the Planck spectrum in generation systems are carried out, the synergistic effects of such conversion are noted, and a map of the conversion modes of solar radiation based on new nanomaterials and nanofluids is presented. In addition, it has is shown that solar multigeneration of electrical and thermal energy can be very effectively combined with methods of desalination and obtaining clean drinking water based on the use of solar radiation concentrators.

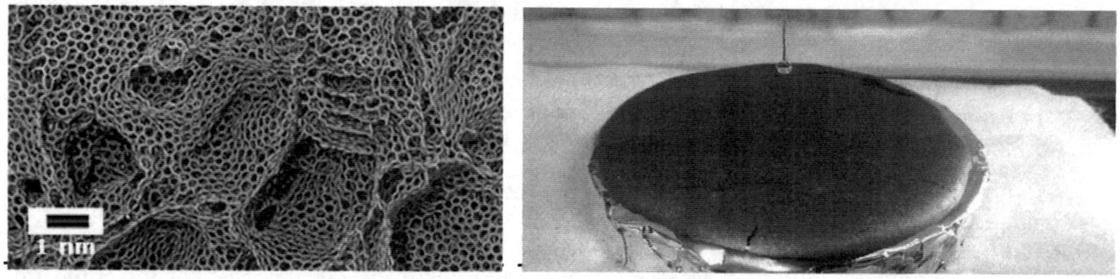

Figure 3. Nanoporous graphene structure [2] (left) and general view of a graphene coating for intense evaporation under the influence of solar radiation (on right)

Figure 4 shows the data of our experiments on the propagation of solar radiation inside containers with various nanofluids. It is easy to see that the most effective absorption is observed in water with graphene nanoflakes (2–3 atomic layers, lateral size of 30–50 μm) [1]. The key mechanism associated with volumetric solar receivers for generating steam is nanoplasmonics - a strong improvement in the absorption of solar radiation, thanks to the mechanism of plasmon resonance. Localized surface plasmon resonance is an optical phenomenon where collective oscillations of free conduction electrons in metal nanoparticles excite radiation with a wavelength comparable to or greater than the particle size [3]. Materials with high conductivity, in particular, metal nanoparticles such as gold, silver, and platinum, are highly efficient centers of absorption and scattering, although they are very expensive for practical use [4].

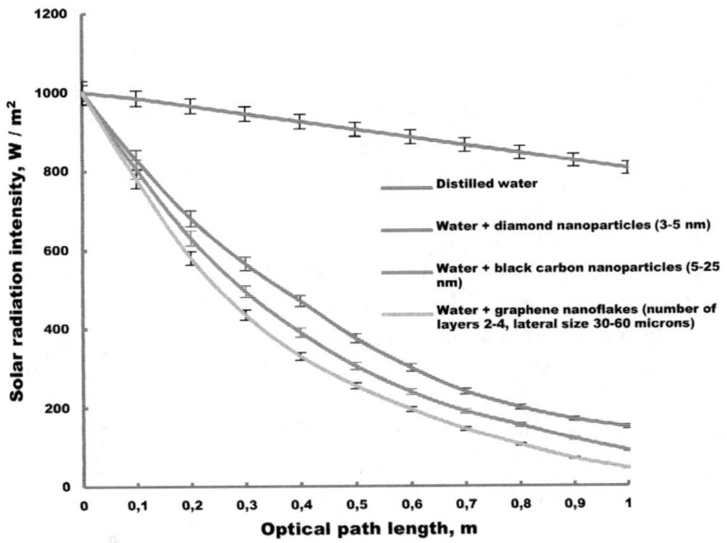

Figure 4. Steam generation processes using nanoplasmonics and nanophotonics [1, 6]

The plasmon resonance wavelength can be controlled depending on the composition, size, geometry, the width of the gap between the nanoparticles and the type of environment.

The generation of vapor upon heating of nanocomponents by solar radiation (nanoparticles of noble metals, nanodiamonds, graphene flakes, etc.), as shown by numerous studies, cannot be fully explained within the framework of the classical mechanisms of bubble nucleation [1,7]. First of all, it should be borne in mind that the local plasmon heating is, as a rule, unsteady; therefore, the temperature of the nanoparticles increases all the time. Second, the nanoparticles are so small that their size is much smaller than the critical diameter of the heterogeneous nucleation of the vapor phase; therefore, the equilibrium mechanism is unlikely. Third, the role of the boundary (interfacial) thermal resistance between a nanoparticle and a liquid, as well as between a nanoparticle and the resulting vapor phase, is completely unclear.

In this regard, it can be argued, based on modern research, that:

1. The mechanism of generation of photothermal bubbles around plasmonic nanoparticles differs from the mechanism of generation of bubbles around microscopic and macroscopic absorbers (including conventional heating surfaces):

• the threshold value of bubble generation under solar irradiation significantly depends on the particle size and duration;

• the mechanism of vapor phase generation during localized heating can be completely different from the mechanism of thermal conductivity during heat transfer from nanoparticles to the environment; at the same time, it may turn out that the heating of the liquid occurs under the influence of a near fluctuating electromagnetic field, the heat flux from which at ultra-small distances from the nanoparticle can be hundreds of times higher than the heat flux, according to the Stefan-Boltzmann relation; this mechanism has not been investigated at all, but the first results have shown its high efficiency;

• an increase in the size of particles and their clustering, in some cases lead to a decrease in the threshold density and initial temperature for steam generation; this fact remains a mystery;

• localized heating of graphene flakes and other highly anisotropic heat sources depends both on the presence of doped nanoparticles inside such nanocomponents (the plasmon frequency and heating efficiencydependence) and on the characteristic dimensions (recall that graphene flakes have a thickness of 3 to 5 nm, and their longitudinal dimensions are tens of microns); the mechanism of the boundary thermal resistance between liquid and graphene flakes and its contribution to the mechanism of vapor generation are still unclear;
• bubbles generated by nanoparticles and nanocomponents can temporarily and spatially localize the thermal field induced by solar radiation and prevent additional heating of working fluids; these collective effects have not been studied experimentally or theoretically, although they play an important role for the efficiency of solar thermal power engineering;
• The mechanisms of floating vapor bubbles - thermal flotation, convective mechanism and their contribution to the efficiency of steam generation - have not been studied at all.
2. Optimization of the conditions for the generation of vapor bubbles can provide a minimum solar heat flux and, consequently, the maximum efficiency and selectivity of plasmon interactions through:
• optimization of the size of nanoparticles and their overall ionic state, in particular, the use of clusters of nanoparticles instead of single particles;
• optimal choice of the size and geometry of the container with the working fluid, if bubbles are generated in closed volumes.
3. The properties of optical scattering of photothermal bubbles provide the possibility of their detection with high sensitivity and time resolution.
Bubbles enhance optical scattering compared to plasmonic nanoparticles. At the same time, this leads to very strong additional scattering of the incident radiation, which requires an optimal choice of the optical thickness of the volume with the liquid.
The problems posed and methods for their solution allow us to take a fresh look at the problems of solar multigeneration, to determine materials, processes and technologies for efficient conversion of solar radiation of the entire Planck spectrum, which will determine further ways of implementing such technologies in engineering and industrial solutions.

2.2. Solar thermoelectric generation
Thermoelectric materials represent a unique opportunity in the field of converting solar radiation into electrical energy without the use of machines. The development of thermoelectric materials over the past decades has taken place in huge leaps, using the latest advances in nanotechnology and nanomaterials: growing semiconductor crystals, controlled doping, nanostructuring and two-dimensional confinement, as well as wave effects in phonon transfer, correlated electronic physics and unconventional transport in organic materials. In addition, the use of 2D materials can be called a promising direction. The search for new nanomaterials for solar thermoelectric conversion is associated with the fact that the figure of merit for them significantly increases due to a sharp decrease in the phonon and electronic thermal conductivity, since they are included in the denominator of the ratio for the figure of merit [8]. In addition, the temperature difference between hot and cold surfaces is essential [9-11].
Since it is the high absorption coefficient of nanomaterials, for example, graphene, that makes it allows obtaining high temperatures of the hot surface, a high temperature difference, up to several hundred degrees may be ensured.
The Seebeck coefficient of the sample increases substantially when that of the grain boundary is high and the thermal conductivity of the grain boundary is low. This can be understood with a simple 1D model

$$S = \{S_\Gamma(1-\delta_{\Gamma B})/\lambda_\Gamma + S_{\Gamma B}(\delta_{\Gamma B}/\lambda_{\Gamma B})\}/\{(1-\delta_{\Gamma B})/\lambda_\Gamma + (\delta_{\Gamma B}/\lambda_{\Gamma B})\}$$

Where S_Γ is the Seebeck coefficient of the grain, $S_{\Gamma B}$ is for the grain boundary, $\delta_{\Gamma B}$ is the thickness of the grain boundary, and λ_Γ *and* $\lambda_{\Gamma B}$ are the thermal conductivities of the grain and the grain boundary.

3. Results and conclusions

Today, a new approach to the selection of materials for solar thermal multigeneration has been formed. Efficient evaporation and vapor generation during surface and volumetric absorption, requires two-dimensional materials, among which the most accessible are graphene and graphene-based composites. This is primarily due to the high nanoporosity (in this case, almost all the pores are open), which forms long open channels of very small diameter - 1–2 nm). The nontrivial hydrodynamics of water in such channels leads to extremely fast movement inside them, the formation of a two-phase liquid-vapor interface with high curvature, and active evaporative heat and mass transfer. This makes allows actively generating steam with a high temperature (in a number of experiments with concentrators, solar radiation reaches a power density of 20 kW / m^2), and the temperature can reach 650 ^{0}C. At the same time, capillary boiling is much more efficient than usual surface boiling, since overheating and steam generation occur inside nanoporous structures.

It has been established for thermoelectric generation that it is very efficient to use high-temperature thermodynamic cycles with a temperature difference between hot and cold surfaces above 200 C (better, up to 400 C). In this case, it is possible to use both silicon-germanium nanocomposites and nanostructures based on bismuth telluride, as well as graphene nanocomposites. To increase the efficiency of thermoelectric conversion, it is necessary to use materials with very low electronic and phonon thermal conductivity. The latter is possible, for example, in new nanomaterials with hydrodynamic electron flow regimes (electron drift motion). In this case, for a sharp decrease in phonon thermal conductivity, it is necessary to use nanomaterials with a small free path (for example, nanopolycrystalline structures with a phonon path length of 3-5 nm). The Seebeck coefficient, power factor, and efficiency, ZT, can be increased by nanostructuring when energy barriers exist around the grain boundaries or embedded nanoparticles.

Acknowledgments

The work is supported by the Russian Science Foundation (Grant No. 17-19-01757).

References

[1] Dmitriev A and Klimenko 2020 Heat power engineering **2**, 1-16
[2] Khan J and Arsalan M 2016 Ren. and Sust. Energy Reviews **55** 414–425
[3] Li X , Min X et al 2020 Joule **2**, 2477-2484
[4] Chang C et al 2020 ACS Omega **4**, 3546–3555
[5] Chang C et al 2019 ACS Appl. Mater. Interfaces **11**, 18466-18474
[6] Dmitriev A S. Hybrid Graphene Nanocomposites: Thermal Interface Materials and Functional Energy Materials In: Graphene Production and Application // Ed. Prof. Sadia Ameen. 2019 InTechOpen
[7] Shi Y, Li R, Jin Y et al Joule **2**, 1171-1186
[8] Cui L, Zhang P, Xiao Y et al 2018 Adv. Mater. 1706805
[9] Chang C, Wang Z et al 2020 Scientific Reports 1**0**: 20500
[10] Pham N, Vallin O and Panda J. 2019 J. Appl. Phys. **125**, 180902
[11] Li Z, Wu Z et al 2020 CCS Chemistry **2(7)** 2174-2182

Improving the compact heat exchangers with profiled tubes for marine power plants

A V Naumenko[1], V D Lychakov[2], M Yu Egorov[2, 3, 4, 5, 6], A A Shcheglov[2], A S Matyash[2], A O Borisov[2] and B F Balunov[2]

[1]Limited Liability Company "Vineta", St. Petersburg
[2]Stock Company "I.I. Polzunov Scientific and Development Association on Research and Design of Power Equipment", St. Petersburg
[3]Peter the Great St. Petersburg Polytechnic University, St. Petersburg
[4]Saint-Petersburg National Research University of Information Technologies, Mechanics and Optics, St. Petersburg
[5]St. Petersburg State University of Aerospace Instrumentation, St. Petersburg
[6]Saint-Petersburg State Forest Technical University, St. Petersburg

mikhail.yu.egorov@gmail.com

Abstract. The design features of heat exchangers with flat tubes are considered. The stages of creating a new flat tube and its prototypes are analyzed. An oil refrigerator is chosen as a prototype of the heat exchanger because of the previous use of flat tubes with wells, now replaced with new flat profiled tubes. Thermal and hydraulic tests of a refrigerator made of such tubes are carried out at the stand. During the testing, the hydraulic resistance of the cavities of the cooled and cooling media and thermal parameters are determined: heat power, heat transfer coefficient and heat transfer coefficient from the cooled medium in the inter-tube cavity with the transverse flow of tubes to the tube wall. A satisfactory correspondence of the actual power values determined for both working environments has been established. The discrepancy does not exceed (-7.6%)-(+5%) with an average value of +0.2%. A satisfactory correspondence of the actual and calculated values of the refrigerator power has been obtained. The discrepancy does not exceed (-15%)-(+9%) with an average value of -2.8%. The calculation of the capacity of the refrigerator during its operation in the design mode of oil cooling is carried out. The oil flow is considered both through the pipe and through the inter-pipe space. A good convergence of the calculations with the experimental results has been revealed.

1. Introduction

The sustainable combat capability of the ship and its survivability are associated with the rational use of areas and volumes to accommodate modern weapons systems and ship equipment. One of the ways of such rationalization is to reduce the size of the ship's power plant and its auxiliary equipment.

The main elements of the auxiliary equipment of the ship's power plant are heat exchangers. Reduction of their masses and dimensions while maintaining or increasing operation efficiency remains an urgent task. This is ensured by intensifying heat transfer processes by turbulizing the flow of working fluid and increasing the effective areas of heat exchange surfaces. To combine the efficiency of plate and shell-and-tube heat exchangers, a new profile of a flat tube of complex cross-section is proposed. Due

Content from this work may be used under the terms of the Creative Commons Attribution 3.0 licence. Any further distribution of this work must maintain attribution to the author(s) and the title of the work, journal citation and DOI.
Published under licence by IOP Publishing Ltd

to this, the weight and size characteristics of the device are minimized while maintaining the required operating modes.

The issues of heat exchangers compactness were approached by such specialists as V.M. Kees, A.L. Longlon [1], as well as several Russian scientists. Since compactness is achieved by intensifying heat transfer, its various methods were considered in the works of G.A. Dreitser and B.V. Dzyubenko [2, 3]. The heat transfer of various surfaces was investigated by such specialists as S.S. Kutateladze [4, 5]. Nevertheless, the issue of compactness of heat exchangers in the SEU requires further study.

The object of research is the technology of manufacturing ship heat exchangers with flat profiled tubes. **The subject of research** is the efficiency and weight and size characteristics of ship heat exchangers. **Purpose of the work:** substantiation of the design and manufacturing technology of compact heat exchangers with profiled tubes for marine technology.

To achieve the set research goal, it is necessary to solve the following tasks:

- to establish the state-of-the-art in the design and manufacturing technology of compact ship-based heat exchangers;
- to develop a model of compact heat exchangers with flat tubes;
- to develop an algorithm and a program for calculating heat exchangers with flat profiled tubes;
- to develop technologies to manufacture heat exchangers in general and a flat tube in particular;
- to establish the effectiveness of flat profiled tubes in ship heat exchangers experimentally;
- to carry out a comparative analysis of the efficiency of heat exchangers with flat and round tubes.

We provide an overview of the main heat exchanger structures, classified according to the method of heat transfer and by purpose. Various heat exchangers are considered by types of structures. The design of heat exchangers with flat tubes is discussed as well. The history of the creation of a new flat tube and its prototypes is briefly described. An MHD-4 oil cooler is chosen as a prototype of the heat exchangers (Figure 1). It previously used flat tubes with dimples, which have been replaced by new flat profiled tubes.

An algorithm and a program for calculating the heat exchangers of the proposed design have been developed. They are based on the thermal calculation of heat exchangers: the outlet temperature of one medium is preset (in this case, the temperature of the medium to be cooled); the balance value of the heat exchanger power is calculated; the outlet temperature of the second medium is determined; the average temperature of both media and the average temperature head are set. Then the values of the heat transfer coefficients of the outer and inner surfaces of the heat transfer pipes and the heat transfer coefficient between the heat exchanging media and the values of the transmitted power are calculated.

2. Heat exchangers with flat tubes

MHD-4 oil cooler is chosen as a prototype of the heat exchanger.

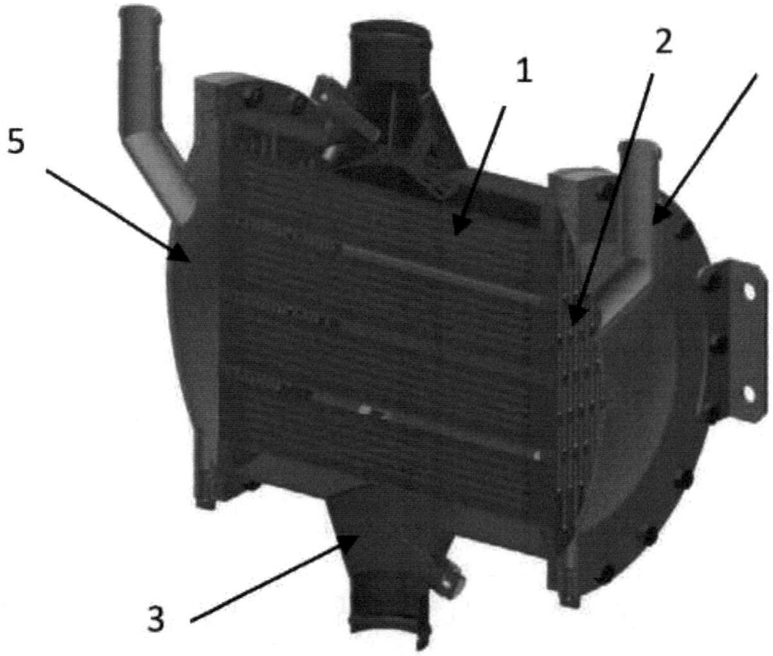

Figure 1. General view of MHD-4.
1 - tube; 2 - tube sheet; 3 - body; 4, 5 - lids

Working fluid in MHD-4 are oil as a cooled fluid and outboard (sea) water as a cooling fluid. The cooled fluid flows in the tubes, and the cooling fluid flows in the annular cavity.

Due to the complexity of the tube shape (Figure 2), the flow in it is uneven, and different velocities lead to different heat transfer coefficients for different parts of the tube.

Figure 2. Sectional profiled flat tube.

To determine the values of the flow rate of the liquid inside, each tube is conventionally divided into 2 elliptical channels and 4 round ones, the calculation of heat transfer for them is carried out separately.

The proposed ratio:

$$\Delta P_i = \left[\left(\frac{\zeta_{in}+\zeta_{tb}L}{dg} + \zeta_{out}\right) G^2/2F^2\rho\right]_i = idem,$$

where

ΔP - hydraulic resistance;

ζ - the coefficient of local resistance;

L – channel length;

dg - hydraulic diameter;

G - the flow rate of the working medium;

F - the area of the passage of the working medium in the channel;

ρ - the density of the working medium;

i - channel number.

In parallel connected channels of two different standard sizes, but of the same length, the flow rates in different channels, the area of the media passage and the coefficients of hydraulic resistance are related by the ratio

$$[K1G/(dg)^2F]_i = idem$$

Then

$$G1/G2=K2(dg1)^2F1/K1(dg2)^2F2,$$

and with a known value of the total flow rate of the fluid through all channels:

$$G1/G\Sigma=1/(1+K2(dg1)^2F1/K1(dg2)^2F2),$$

and

$$G1= G\Sigma\text{-}G2.$$

It is possible to find the flow rate of the working fluid in each channel, and, subsequently, the speed of the working fluid, the Reynolds number and the heat transfer coefficient. The final heat transfer coefficient was calculated as an average, considering the perimeters of the channels.

The transverse flow around the tube bundle is considered by analogy with the flow around the in-line bundle of membrane tubes.

To calculate the flow around a bundle of ribbed tubes in general form, the heat transfer coefficient from the flow to the tube wall is first determined when flowing around a similar bundle of smooth tubes. Then, to consider the effect of ribbing, the heat transfer coefficient for a bundle of smooth pipes is multiplied by a correction coefficient.

Thermal-hydraulic tests of the oil cooler MHD-4 were carried out at the complex stand of NPO CKTI (St. Petersburg). To simplify the testing, the TA working media were replaced with fresh water in both circuits. The features of the bench equipment regulated the supply of the cooled fluid into the inter-tubular cavity of the heat exchangers, and the cooling one directly into the tubes.

In the process of testing the MHD-4 refrigerator, the hydraulic resistance of the cavities of the cooled and cooling fluid and the thermal parameters (thermal power, heat transfer coefficient and the coefficient of heat transfer from the cooled medium in the annular cavity with transverse flow around the tubes to the tube wall) were determined (Table 1). Conclusions have been drawn on

- satisfactory correspondence of the actual power values, determined for both working environments

(Mi; M2). The discrepancy does not exceed (-7.6%) - (+ 5%) with an average value + 0.2%;

- satisfactory correspondence between actual and calculated power values MHD-4 (No. and M).

The discrepancy does not exceed (-15%) - (+ 9%) with an average of -2.8%.

Table 1 - Test results of MHD-4.

№	Cooled fluid						Cooling fluid						Np
	G1	P1	Tin1	Tout1	Nd1	dP1	G2	P2	Tin2	Tout2	Nd2	dP2	
	kg/s	MPa	°C	°C	kWt	kPa	kg/s	MPa	°C	°C	kWt	kPa	kWt
1	2.39	0.19	28.5	25.5	29	3.71	0.8	0.37	2.9	11.1	27	2.00	33
2	2.05	0.20	31.3	27.6	32	3.43	0.79	0.37	2.9	12	30	2.02	35
3	2.21	0.19	42.6	37.6	47	3.5	0.78	0.36	2.8	16.7	46	1.97	50
4	2.52	0.17	45	40.2	50	3.84	0.77	0.36	2.8	18	49	1.84	54
5	2.18	0.19	49.4	43.3	55	3.6	0.78	0.36	2.7	19.3	54	1.92	59
6	2.47	0.17	50.8	54.3	57	3.97	0.77	0.36	2.7	20.3	57	1.94	62
7	2.69	0.17	52.6	47.3	60	4.18	0.77	0.36	2.8	21.2	60	1.9	65
8	2.48	0.18	54.4	47	76	3.94	1.43	0.33	2.7	15.4	76	6.25	82
9	2.18	0.20	60.7	53	70	3.85	0.77	0.35	2.8	24.5	70	1.91	73

10	2.44	0.18	63.1	54	94	4.1	1.56	0.32	2.7	17.2	95	7.16	98
11	1.85	0.23	69.7	59.3	80	3.48	0.76	0.36	2.8	28.5	82	1.82	82
12	2.55	0.18	70.6	62.3	90	4.41	0.75	0.35	2.8	29.5	84	1.89	87
13	2.52	0.17	71	60.1	115	4.25	1.78	0.3	2.7	18.8	120	9.57	116
14	2.17	0.22	72.3	63.5	80	3.73	0.57	0.35	2.7	35.8	79	1.07	78
15	2.13	0.21	79	70.2	78	4	0.46	0.36	2.7	42.9	78	0.7	80
16	2.11	0.21	80.4	69.4	97	4.07	0.8	0.35	2.6	32.6	101	2.07	100
17	2.46	0.18	81	71.1	102	4.31	0.8	0.35	2.5	33.6	104	1.97	103
18	2.54	0.18	81.3	68	141	4.2	1.8	0.29	2.4	21.7	146	9.66	134
19	2.27	0.21	82	75.5	62	4.1	0.27	0.36	2.5	56.1	61	0.03	69
20	2.07	0.23	89.4	81.8	66	4.23	0.27	0.36	2.6	61	67	0.23	75
21	2.07	0.22	91	78.1	112	4.18	0.81	0.34	2.4	37	117	2.09	114
22	2.38	0.2	91.2	79.5	117	4.16	0.81	0.34	2.4	37.8	119	1.97	117
23	2.5	0.29	91.4	76	161	4.18	1.77	0.3	2.3	24.8	167	9.53	150
24	2.14	0.23	98.8	90.6	73	4.2	0.27	0.36	2.8	68.5	75	0.15	83
25	2.12	0.23	99.9	86	124	4.16	0.81	0.36	2.5	41.1	131	1.85	127
26	2.39	0.21	99.8	86.8	131	4.33	0.81	0.36	2.5	41.6	133	2.12	128
27	2.46	0.2	99.6	82.5	177	4.26	1.7	0.32	2.4	27.5	179	8.86	162
28	2.38	0.21	99.5	88.5	110	4.16	0.57	0.37	2.5	50	113	1.03	113
29	5.5	0.2	69.5	64.1	124	21.5	1.66	0.36	2.5	20.4	125	8.43	24
30	6.97	0.18	69.8	65.2	134	33	1.68	0.33	2.4	21	131	8.65	128
31	8.35	0.18	70.5	66.7	133	47.1	1.7	0.36	2.3	21.1	134	8.81	132
32	11.2	0.21	70.1	67.1	141	86	1.65	0.34	2.3	22.1	137	8.41	132

3. Heat exchangers with flat tubes of other designs

Replacing round tubes with flat ones in shell-and-tube coolers of the OKP series allowed reducing the weight and size characteristics of existing coolers 2-2.5 times.

The oil is the cooling fluid flowing in the vessel, and the cooling medium (seawater) flows in the pipes.

The OKP 29-42 cooler is selected as a prototype (Figure 3).

Figure 3. General view of OKP 29-420.

The purpose of the tests was to establish the agreement between the measured thermohydraulic characteristics of the modified heat exchanger and the calculated data, to compare them with the characteristics of a shell-and-tube cooler of a classical design with round pipes.

The algorithm and program for the thermal calculation of the OKP 29-420 cooler are built on the same principle as for the MHD-4 oil cooler, but they have several differences. So, to calculate heat transfer at the outer surface of pipes in the ideal case (no leaks through the baffles, no additional baffles, only the transverse flow around the tube bundle is considered), recommendations for the calculation of heat transfer on the surface of the located corridor membrane bundles were used. For the design operating mode of the OKP 29-420 cooler (Tout = 35° C and N = 702 kW), the temperature of the oil at the outlet of the cooler (T1out = 87° C) and its power (N = 313 kW) were determined. The calculation results are shown in table 2:

Table 2. Comparison of project and calculated data OKP 29-426.

Initial data				Required values for the project		Calculation results		
G1	T1in	G2	T2in	T1out	N	N1out	T2out	N
kg/s	°C	kg/s	°C	°C	kWt	°C	°C	kWt

2.7	120	11.1	20	35	702	87	26.5	313

Due to the revealed inconsistency between the design and calculated data, the formula for calculating the thermal conductivity coefficient is used:

$$\alpha = 10{,}5\lambda Pr^{0{,}33}(\frac{G}{\mu})^{0{,}59}$$

The actual value of the heat transfer coefficient of the cooled fluid α_l is three times less than the calculated value of α_{theor}. This is due to the presence of medium flows through the gaps in the partitions between the channels and the stagnation of the fluid in the cavities between the additional baffles.

To reduce the area of the gaps between the intermediate baffles, the design of the profiled tube is changed. In this case, the pipe sections passing through the baffles are not profiled (see Figure 4).

In the old design, the tubes were profiled along the entire length, with the exception of the end areas of connection to tube sheets (see Figure 5).

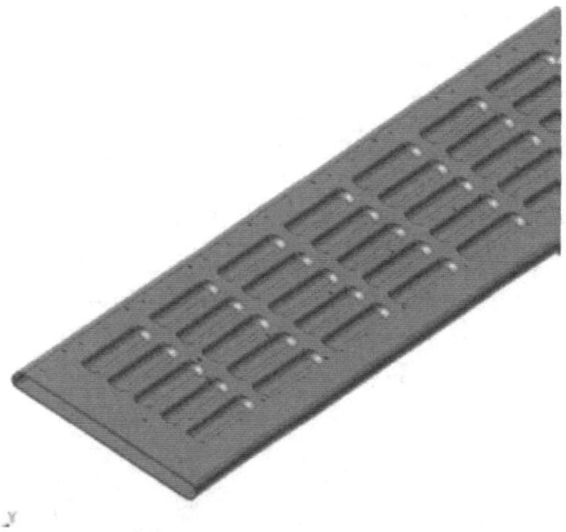

Figure 4. New tube design.

Figure 5. Old tube design.

To ensure the counterflow of working fluid inside the heating unit, a longitudinal baffle is introduced (Figure 6). The proposed design has no seal assembly, but the technological gap between the longitudinal baffle of the tube bundle and the shell is minimized. To ensure a uniform flow of the working fluid in the annular cavity, the tube bundle is rotated 90 degrees.

Negative factors are minimized. The heat exchange surface area has increased by 20% without increasing the body diameter.

Table 3. Calculated data for various cooler designs.

T_{1in}, °C	T_{1out}, °C	N,kWt
	Without longitudal baffle	
120	45.4	635
80	37.4	314
	With longitudal baffle without seal unit	
120	42	657
80	35.2	328
	With longitudal baffle and seal assembly	
120	35.5	691
80	30.8	349

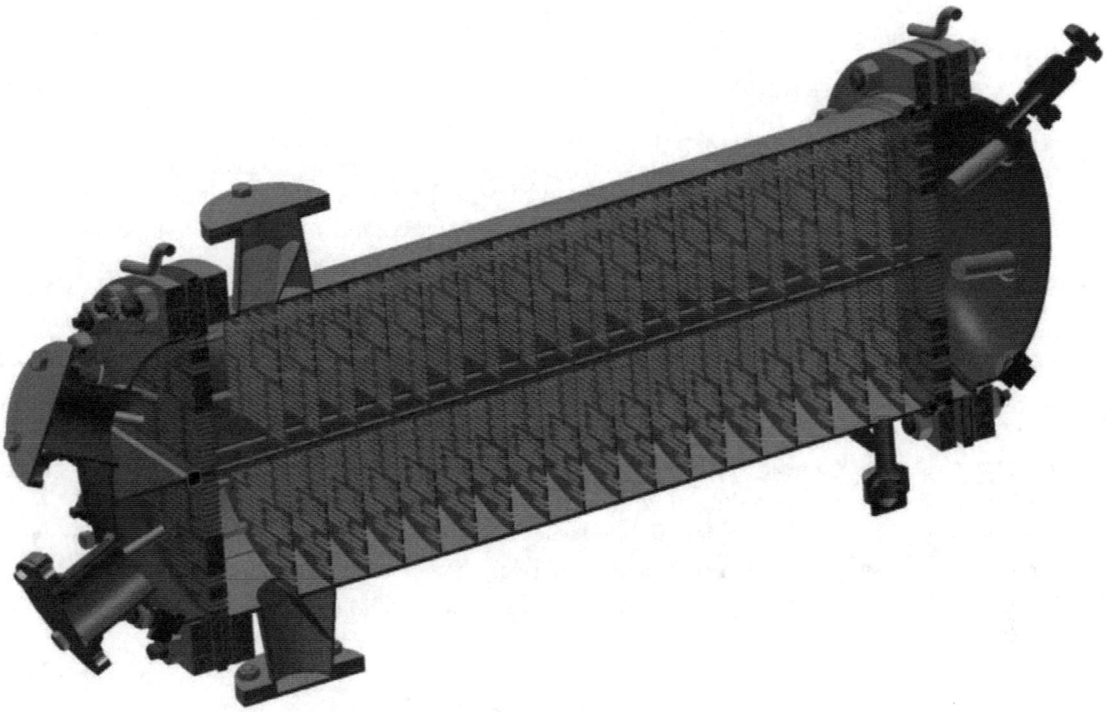

Figure 6. General view of OKP 29-420.

The OKP 29-420 cooler of a new design with flat tubes and a cooler with round tubes have passed the heat engineering tests organized by Vineta LLC with Azrogaz LLC on the territory of the CIAM Research Center (Lytkarino, Moscow region). It has been established that the characteristics of flat-tube and circular-tube coolers are fully consistent when operating in accordance with the specification requirements (Table 4 and 5). According to the test results, flat-tube coolers of the OKP type are recommended for use on various ships.

Table 4. Test results of circular tube cooler OKP 29-420 at project operating modes.

	Temperature	≈50		≈80		≈120	
Heating	Oil inlet/outlet	50/31.2	50.8/31.9	80.8/37.6	82.3/39.2	117/48.7	120/50.3
	Water inlet/outlet	26/27.5	26/28	26/31	27/32	28/35	28/36
Cooling	Oil inlet/outlet	50/35.2	51.6/36	80.8/42.4	83.2/43.1	108/49.5	111.1/50.3
	Water inlet/outlet	30/32	30/32	30/33	29/33	28/35	28/35

Table 5. Test results of flat-tube cooler OKP 29-420 at project operating modes.

	Temperature	≈50		≈80		≈120	
Heating	Oil inlet/outlet	50/28.7	50.8/29.5	80/38.4	81.5/39.2	120/51.1	120.5/51.8
	Water inlet/outlet	16/18	17/19	17/22	17/22	17/25	17/25

Cooling	Oil inlet/outlet	50.8/31.2	53.1/31.9	80.8/40.8	81.5/41.6	111/50.3	117/51.8
	Water inlet/outlet	19/21	19/21	19/23	19/23	18/25	18/25

4. Conclusion

A model of a new shell-and-tube heat exchanger with flat profiled tubes is proposed. It differs from shell-and-tube heat exchanger with round tubes by greater compactness while maintaining efficiency.

To eliminate the discrepancy between the calculation and experiment results, the dependence is introduced into the model to determine the heat transfer coefficient when flowing around a bundle of flat tubes.

The study has shown that the sustainable combat capability of the ship and its survivability may be ensured by reducing the size of the ship's power plant and its auxiliary equipment. Intensification of heat transfer in a heat exchanger due to turbulization of the flow of working media and an increase in the effective areas of heat exchange surfaces allows reducing the mass and size characteristics of devices 2-2.5 times while maintaining the design characteristics.

The proposed model, algorithm and program for calculating heat exchange with flat profiled tubes serve to reach a satisfactory correspondence between the actual and calculated values of the power of MHD-4 and OKP 29-420. The discrepancy ranges from (- 15%) to (+ 9%) with an average of -2.8%.

References

[1] Kays W, London A 1998 *Compact heat exchangers*
[2] Dzyubenko B, Kuzma-Kitscha Yu, Leontiev A, Fedik I 2017 *Intensification of Heat and Mass Transfer on Macro-, Micro-, and Nanoscales*
[3] Vilemas Yu, Voronin G, Dzyubenko B 1988 *Heat transfer intensification*
[4] Kutateladze S, Borishansky V 1958 *Handbook of heat transfer*
[5] Kutateladze S 1990 *Heat transfer and hydrodynamic resistance*

Active thermal interface graphene nanocomposites for thermal control of electronic and power devices

D D Babenko, A S Dmitriev and I A Mikhailova

National Research University «MPEI», Russia, 111250 Moscow, 14 Krasnokazarmennaya Street

asdmitriev@mail.ru

Abstract. New experimental and calculated data are presented for active thermal interface materials, in which heat is removed not only due to high thermal conductivity, but also due to the evaporation of liquids, for example, water, inside a nanoporous graphene structure. It is shown that such active thermal interfaces may be new systems of active thermal control.

1. Introduction

The development of modern and promising electronic and energy devices requires ever more efficient thermal control systems for thermal stabilization and management of thermal conditions [1,2]. Until now, thermal control systems have been divided into two broad categories. The first is thermal interface materials (TIM), i.e., passive materials to reduce the thermal boundary resistance between different blocks of devices, for example, a processor and a heat sink (e.g., a radiator). The second is active cooling systems (based on convective or evaporative cooling, for example, evaporative heat pipes). However, recent developments of new nanomaterials, including graphene, graphene nanocomposites and other two-dimensional materials (boron nitride, aluminum nitride, molybdenum disulfide, etc.) allow developing and creating active thermal interface materials (active TIM), which simultaneously reduce thermal boundary resistance (Kapitza resistance at interfaces of various materials), and cool such interfaces due to either microchannel or evaporative cooling [1]. A significant step towards the creation of active TIM was the discovery of non-trivial capillary hydrodynamics of working fluids, in particular, water, inside porous graphene nanocomposites, which enables their quick delivery to the evaporation surface through the quasi-laminar structure of graphene nanoclaps [2-6].

In this work, we experimentally and theoretically investigate the main processes in active TIM, including thermohydrodynamics of flow and heat transfer inside quasi-laminar structures of graphene nanoflakes with different numbers of monoatomic thickness of graphene plates and study the processes of surface evaporation from graphene nanocomposites in a wide range of temperatures from room temperature up to Leidenfrost temperature and above. It is shown that the fast flow of water and other working fluids is associated with nontrivial capillary hydrodynamics of the non-Poiseuille type, and heat transfer inside such structures is associated with efficient heat transfer (low boundary thermal resistance between graphene nanoflatter and fluid flows due to a sharp increase in the velocity of fluid sliding along the surface - the Navier effect in the hydrodynamics of nanostructures [1]). The processes of wetting and evaporation of droplets of working fluids are investigated in different temperature ranges, and the dependences of the droplet evaporation rates on

Content from this work may be used under the terms of the Creative Commons Attribution 3.0 licence. Any further distribution of this work must maintain attribution to the author(s) and the title of the work, journal citation and DOI.
Published under licence by IOP Publishing Ltd

the heterogeneity and roughness of the surface are shown, which is important for the correct choice of technologies for the manufacture of new materials for active TIM.

2. Materials and method

The experimental cell consisted of a silicon heated substrate, a container inside which an evaporating liquid was supplied (this work presents data only for water and aqueous graphene nanofluids), a graphene structure block placed inside steel meshes of copper or stainless steel, and consisting of a nanoporous structure steam condensation unit. The condenser could force condensed liquid into the cooling unit (or similar to heat pipes) (Figure 1).

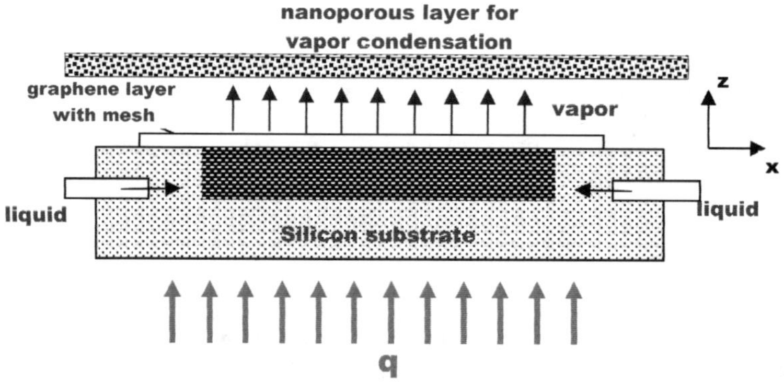

Figure 1. Schematic of an experimental nanoporous graphene evaporation cell

Graphene nanoflakes with an average size of 10-30 μm (thickness of 1.5-3.5 nm) were received from Graphene Institute Ltd (Russia). Chemically clean graphite microparticles of ~300 μm were used. Cooper M1(Cu>99.6%) or stainless-steel mesh with cell side size of 0.3 mm and wire diameter of 0.1 mm supplied by TDMC, Russia, were used as a mesh. A nanoporous oriented layer of graphene nanoflakes before its placement inside the experimental cell is shown in Figure 2. Graphene nanoporous oriented structures were fabricated by the original "domino method" - the stacking of graphene flakes with a flow of rotating liquid through a narrow slit with a gap of about 1 mm.[7] Graphene nanoflakes had a thickness of 1.5-3.5 nm and lateral size from 20 to 100 microns, and a substrate was made of these nanoflakes for active thermal control (high thermal conductivity and highly efficient evaporative cooling through nanopores). Photo and electron microscopy of the samples is shown in Figure 3. The temperature of the silicon substrate varied from 25 to 65 ^{0}C. It is clear that the evaporation rates and cooling efficiency increased with an increase in the substrate temperature. The substrate temperature was measured with thermocouples and heat flux sensors. The temperature was measured with an accuracy of about 0.5 ^{0}C. The evaporation rate was measured by an accurate electronic balance inside the nanoporous capacitor, as well as by the liquid flow rate through the heating unit.

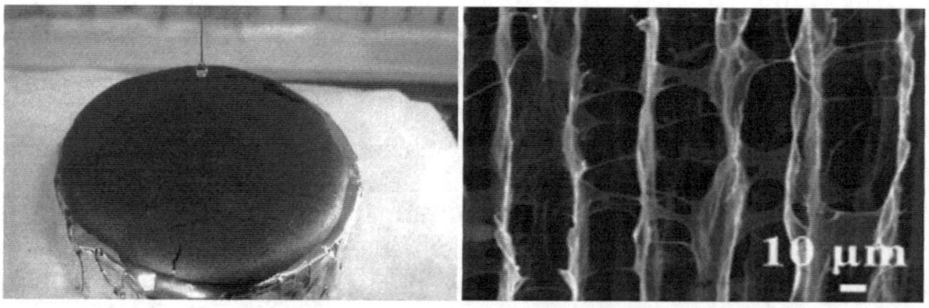

Figure 2. Graphene structure (wick) with mesh for placement inside the evaporation unit

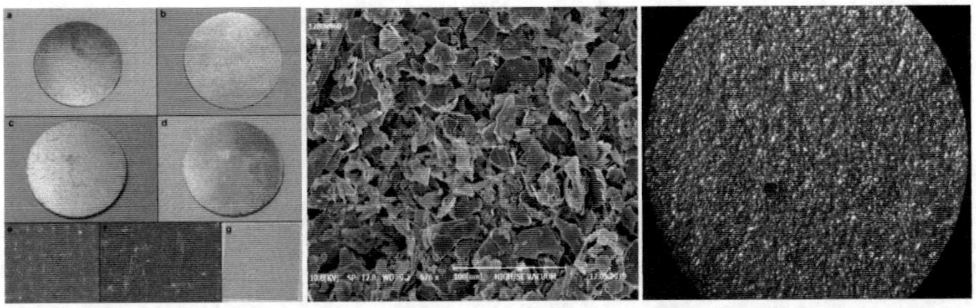

Figure 3. Photos and microscopy of samples (left): a) graphene nanoflakes p=300 bar; b) graphene nanoflakes with cooper mesh p=300 bar; c) graphene nanoflakes with cooper mesh p=180 bar; d) graphite microparticles with cooper mesh p=300 bar; e) sample B optical microcopy x4 magnification; f) sample B optical microcopy x10 magnification; g) sample of cooper mesh 03x01. Graphene nanoflakes (thickness of 1.5-3.5 nm, lateral size from 20 to 100 microns) and a substrate made of them for active thermal control (high thermal conductivity and highly efficient evaporative cooling through nanopores) (center and right)

3. Cooling the model chip using active thermal interface material

In our work, to observe the cooling rate of the model chip, in addition to contactless thermometry and a set of thermocouples, a high-speed high-resolution infrared camera Artcam-320 is used. An infrared thermogram is shown in figure 4. It is clearly seen that from an initial temperature of 55 ° C due to evaporative cooling of the nanoporous graphene structure after about 40 seconds, the substrate is cooled to 28 ° C. The cooling rate is thus about 0.675 degrees per second. This means that due to the removal of steam to the external environment or to the receiver, a heat flux of more than 120 W / m^2 can be removed.

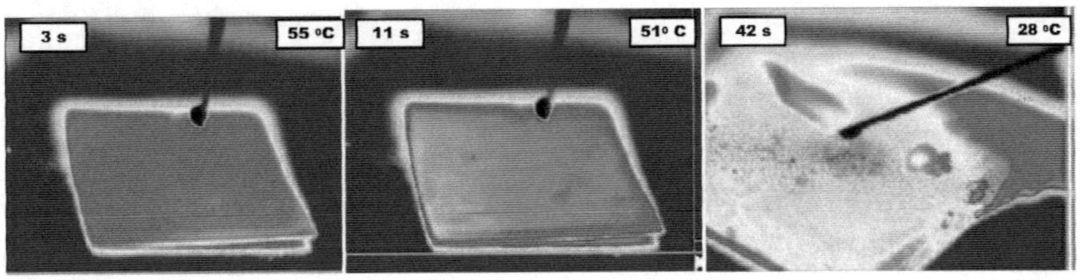

Figure 4. Infrared thermograms of the model chip cooling when using an active thermal interface material

4. Evaporation rate data

In addition to pure liquids, we were the first to study the evaporation of graphene nanofluids inside nanoporous structures. We investigated both the evaporation of pure liquids, in particular distilled water, and nanofluids based on graphene nanoflakes of various lateral sizes (from 10 μm to 60 μm). Naturally, the evaporation of graphene nanofluids occurred inside structures with large pores (from 10 to 100 μm). It was quite surprising that graphene nanofluids penetrated through the graphene structure with negligible resistance, practically forming a denser layer in the upper part. As a result of the experiments, it was found that when the concentration of graphene in the nanofluid was no more than 1%, there was practically no decrease in the evaporation rate. Preliminary data demonstrated that such fluids heat up and evaporate more efficiently, as shown by the empirical data in Figure 5.

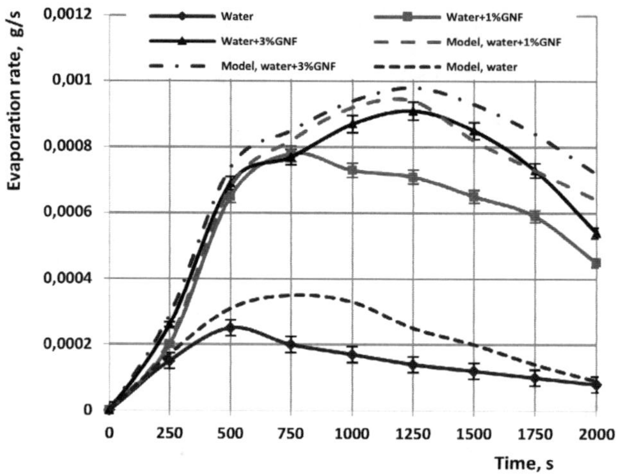

Figure 5. Evaporation rates of various liquids using graphene nanochannels and nanopores

The physical mechanisms of the increased evaporation rate are not yet completely clear and require additional study. At the same time, the use of graphene nanofluids together with graphene nanoporous structures allow creating an active system of evaporative cooling of a continuous cycle, which is extremely important for a large number of applications. The obtained experimental data on the evaporation of water and aqueous graphene nanosuspensions with mass concentrations of 1% and 3% are shown in Figure 5.

5. Mathematical models

A feature of evaporation from oriented graphene nanostructures is nontrivial capillary hydrodynamics - a strong influence of the boundary conditions on the liquid - graphene plate surface on the movement of the meniscus of the interface (almost complete water sliding without tangential resistance - (Navier boundary condition and significant slip length) [4]. In this work, we used models close to those developed in the [5].

5.1 Capillary-driven flow

The imbibition of liquid into a nanoporous structure occurs when the solid–vapor specific surface energy (σ_{sv}) is greater than the solid–liquid specific surface energy (σ_{sl}). This usually occurs when the contact angle is less than 90^0. The driving force for imbibition is characterized by the Young–Laplace equation

$$\Delta p = p_l - p_v = 2\sigma_{lv}\Gamma$$

where Δp is the local capillary pressure, p_l is the local liquid pressure at the liquid–vapor interface, p_v is the pressure of the vapor phase, σ_{lv} is the liquid–vapor surface tension, and Γ is the local mean curvature of the liquid surface.

In contrast to work [5], where the Brinkman equation (or an analogue, the Darcy equation) was used for the flow of a liquid inside a porous structure, in this study we describe the flow of a liquid inside a graphene structure based on non-slip flow. In this case, the slip length significantly exceeded the characteristic scale of nanopores $L_s >> \delta$ ($L_s \sim 50$ nm, $\delta \sim 3$-5 nm).

5.2 Heat transfer

Heat in thermal control devices is generated due to Joule heating and is transferred to the liquid vapor interface through the combined solid-liquid matrix by the heat conduction mechanism. Further, heat is dissipated by convection, heat conduction and thermal radiation into the environment. Typically, thermal control systems use a heat transfer coefficient $\alpha=q(T_s-T_\infty)$, q is the steady state heat flux dissipated by the graphene nanostructures (wick), T_s is the temperature at the solid interface, and T_∞ is the temperature of the far ambient field. The effective thermal conductivity of graphene nanostructures is dependent on the working fluid, the solid material and nanostructures and operating conditions. In the simplest case, the following model can be adopted (the study shows that the complication of the model insignificantly affects the final result)

$$\lambda_{eff} = \lambda_s \frac{2+\lambda_l/\lambda_s - 2\phi(1-\lambda_l/\lambda_s)}{2+\lambda_l/\lambda_s + \phi(1-\lambda_l/\lambda_s)}$$

where λ_s, λ_l are the thermal conductivities of the solid and liquid, and ϕ is the porosity. When using nanofluids, the quantity is replaced by the effective thermal conductivity of the nanofluid ($\lambda_l \rightarrow \lambda_{l,eff}$).

5.3 Vapor flow

The mass flow of vapor during evaporation can be determined by the following relationship

$$\vec{j}_v = \rho_m \varphi \vec{u}_v - \rho_m D_v \nabla \varphi$$

where j_v is the vapor mass flux vector, ρ_m is the mixture (air + vapor) density, φ is the vapor mass fraction, and D_v is the diffusivity of vapor in the gas (or mixture of gases). The evaporative mass flux is related to the evaporative heat flux by $q_v = j_v \alpha_{lv}$ equivalent thermal resistance is $R_v = (T_s - T_\infty)/j_v \alpha_{lv} S$.

Using the above model, the rates of evaporation of various liquids (water and water suspensions of graphene) have been calculated. These calculated data are presented in Figure 5, which also shows experimental data in the study of evaporation from graphene oriented nanostructures.

Unfortunately, within the framework of the used model, there is no good agreement between theoretical and experimental data, which, first of all, indicates the shortcomings of the existing models of evaporation from nanostructures. Note that the results of the calculations give data exceeding the evaporation rate found experimentally. It should be especially emphasized that for the case of liquid sliding inside graphene nanochannels and evaporation from the surface of the meniscus inside the channel, an interesting situation may arise - the liquid inside the channels can move much faster (slip boundary condition) than the liquid-vapor interface. As far as we know, such cases have not been previously considered within the framework of mathematical models.

In our opinion, it is precisely in these processes that the nontrivial capillary hydrodynamics of liquids inside the nanochannels manifests itself. Therefore, in addition to the importance of further constructing models of vapor transport in the external environment, it is necessary to more consistently consider capillary nanohydrodynamics inside channels, especially when it concerns two-dimensional materials such as graphene and its structure.

5.4. Cooling

The fraction of heat dissipated by evaporation was determined by subtracting the perceptible cooling and parasitic heat loss from the heating source. The heat flux used for cooling was calculated from the ratio of the energy balance

$$q_c = \frac{[j_v + j_L] C_p (T_{out} - T_{in})}{S_{Cool}}$$

where q_c is the heat flux for cooling, j_v is the vapor mass flux, j_L is the evaporation loss, C_p is the specific heat capacity of the working fluid, and S_{Cool} is the cooling surface area. The table 1 shows the calculation results and experimental data on the dependence of heat fluxes during cooling on the type of working fluids and the heat transfer coefficient.

Table 1. The dependence of heat fluxes

Working fluids	α, W/cm^{2}K	Maximum heat flux, W/cm^2	Publication
Water	9.6	140	[6,8]
Water	14.8	180	[6,8]
Water	38	390	[6,8]
R-245fa	39.7	550	[6,8]
Pentane	18.4 (th – 20.2)	230 (theory – 244)	[this work]
Water	24.7 (th – 20.2)	297 (th –316)	[this work]
Water+GNF1%	31.2(th – 34.8)	340 (th – 369)	[this work]
Water+GNF3%	34.7(th – 34.8)	365 (th – 369)	[this work]

6. Conclusions

As a result of experimental studies and on the basis of the developed mathematical models we have revealed active thermal interfaces, in which, in addition to high thermal conductivity, heat removal is carried out by evaporative cooling of graphene nanoporous structures. In addition to ordinary liquids (for example, water), nanofluids, including graphene nanocolloids, can also be used. It has been established that in the future such active TIMs can replace passive TIMs and provide the necessary thermal control of energy and electronic devices.

Acknowledgments
The work is supported by the Russian Science Foundation (Grant No. 17-19-01757).

References

[1] Alexander S. Dmitriev 2019 Hybrid Graphene Nanocomposites: Thermal Interface Materials and Functional Energy Materials, Graphene Production and Application, Sadia Ameen, M. Shaheer Akhtar and Hyung-Shik Shin, *IntechOpen*
[2] Singh H and Myong R Adv. Mat. Sci. and Eng. **2018** 9565240
[3] Zhu Y, Zhan K and Hou X ACS Nano 2018 12 908–911
[4] Lei W, Rigozzi M and McKenzie D Rep. Prog. Phys. 2016 **79** 025901
[5] Wilke K, Barabadi B, Lu Z, Zhang T and Wang E Appl. Phys. Lett. 2017 **111** 171603
[6] Vaartstra G, Zhang L, Lu Z et al J. Appl. Phys. 2020 **128** 130901
[7] Kairi M, Dayou S, Kairi N et al J. Mater. Chem. A, 2018, **6**, 15010
[8] Hanks D, Lu Z, Sircar J et al ACS Appl. Mater. Interfaces 2020 **12**, 7232–7238

Comparative analysis of characteristics of liquid fuel combustion for various designs of oil-steam burners

I S Anufriev and E P Kopyev

Kutateladze Institute of Thermophysics SB RAS, Novosibirsk, Russia

kopyeve@itp.nsc.ru

Abstract. The paper presents a comparative analysis of experimental data on combustion of liquid hydrocarbon fuels in the presence of superheated steam in two designs of the developed burners. By the example of diesel fuel burnt in a spray burner it is shown that lower values of nitrogen oxide contents in the exhaust gases are achieved in comparison with an evaporative burner. At that, the content of carbon monoxide in some regimes is lower for the evaporative burner. The regimes with the minimum content of toxic combustion products are found for both designs of the burners.

1. Introduction

Currently, the issue of environmentally friendly combustion of heavy hydrocarbons and substandard fuels for cheap energy production is relevant [1,2]. One of the promising approaches to reduce toxic products when burning liquid combustible waste is steam injection into the combustion zone as an effective way to suppress nitrogen oxides and soot formation [3,4].

Water and steam are used in practice to solve a number of problems: from reducing the heat load and suppressing detonation [5] in the combustion chamber to dispersing heavy fuels and burning water-fuel emulsions. The addition of steam both changes the process of some elementary reactions (chemical effect) and leads to dilution of the combustible mixture (physical effect), which ensures a decrease in the flame temperature [6]. The presence of steam in the combustion zone leads to a decrease in NOx emissions due to an increase in the heat capacity of the mixture during combustion of hydrogen-air [7] and methane-air [8–10] mixtures, as well as natural gas [11].

However, the efficiency of using water or steam injection greatly depends on various factors: type of device, method of supply, type of fuel, steam concentration, etc. In gas turbines using natural gas [3], steam injection into the combustion chamber is the most effective for NOx suppression. And when burning kerosene, it is more useful to inject steam into the ignition area [12]. The best effect is achieved when steam is added to the region of the maximum flame temperature, [13] which, in particular, can be located at the end of the combustion zone [14].

Despite the abundance of works dealing with the effect of steam and water on the combustion process, there is very little information in the literature on the combustion of substandard liquid fuels (such as waste oils, refinery waste, crude oil, etc.) with steam injection and the effect of steam parameters and methods of supplying steam and fuel on combustion indicators. Therefore, to solve the problem of environmentally friendly disposal of liquid combustible waste, the authors have previously studied two new designs of burners with fundamentally different methods of supplying superheated

Content from this work may be used under the terms of the Creative Commons Attribution 3.0 licence. Any further distribution of this work must maintain attribution to the author(s) and the title of the work, journal citation and DOI.
Published under licence by IOP Publishing Ltd

steam and fuel and their mixing, developed at IT SB RAS, using the example of burning diesel fuel. These are laboratory samples of an original vertical evaporative burner with controlled forced supply of superheated steam to the combustion zone [15] and an atmospheric burner with spraying liquid hydrocarbon fuel by a jet of superheated steam [16]. It is shown using the evaporative burner [15] that when fuel is burned in a jet of superheated steam, the content of nitrogen oxides and carbon monoxide in the combustion products is significantly lower than the maximum permissible concentrations established for this type of burner, according to the performance standards [17]. The developed spray burner [16] demonstrates the possibility of efficient combustion of diesel fuel and waste engine oil [18]; a 30% reduction in NOx was ensured with the supply of steam in comparison with air due to the effect of diluting the combustible mixture and lowering the flame temperature.

The purpose of this work is a comparative analysis of the experimental data obtained earlier by the authors during the combustion of liquid hydrocarbon fuels with addition of superheated steam in two different original designs of laboratory burners aimed at determining the most preferable method of steam supply to the combustion zone.

2. Burners

2.1. Evaporative burner with forced supply of superheated steam

An evaporative burner with forced supply of superheated steam was studied by the authors in [15]. A detailed description of the burner, the methods for measuring its characteristics and the results obtained in the presence of a supplied jet of superheated steam are presented in [15]. The experiments were also carried out with the supply of a jet of heated air instead of steam; their results are given in [19]. The principle of burner operation is based on the combustion method proposed by the authors of [20], when the products of thermal decomposition and incomplete combustion of evaporating liquid fuel are gasified at a high concentration of steam in the combustion chamber, and the reaction products are burned out in an external flame (Fig. 1). Since the burner is of an evaporative type, its power is limited by the rate of fuel evaporation, including limitation due to the furnace size. The optimum fuel flow rate was previously defined as 0.8 kg/h for diesel fuel. The burner design provides for a natural flow of atmospheric air into the reaction zone through the openings in the lower part of the combustion chamber. This air flow is required to ignite evaporating liquid fuel in the combustion chamber. Two steam nozzles connected to an external steam generator (opening angle of 17°) are installed coaxially above the combustion chamber at the base of the gas generation chamber and are oriented vertically (outlet diameter of 0.5 mm). The lower nozzle is directed downward into the prechamber (with an inner cavity in the form of a hemisphere and a conical outlet). A steam jet is supplied to the combustion chamber to remove thermal decomposition products, which prevents coking of the inner surfaces. The upper nozzle is directed upwards into the gas generation chamber.

2.2. Spray burner

Characteristics of the spray burner operation both in the case of supply of a high-speed jet of superheated steam and a jet of heated air instead were investigated in [16]. The main feature of the device is organization of an original method for dispersing liquid fuel (Fig. 2): a high-speed jet of superheated water vapor is fed into the gas generation chamber through a steam nozzle (opening angle of 17° and outlet diameter of 0.5 mm) installed at the burner base and directed vertically upward. Liquid fuel flows through the fuel supply pipe into the steam jet base at a predetermined flow rate; as a result, a homogeneous fine-dispersed gas-droplet flow is formed. In addition to spraying the fuel, superheated steam raises the temperature of fuel droplets, which intensifies mass transfer and mixture formation, contributing to stable ignition. The advantage of using this scheme over an evaporative burner is that when spraying fuel, it is possible to vary the power of the burner over the wide ranges.

Journal of Physics: Conference Series **2150** (2022) 012009 doi:10.1088/1742-6596/2150/1/012009

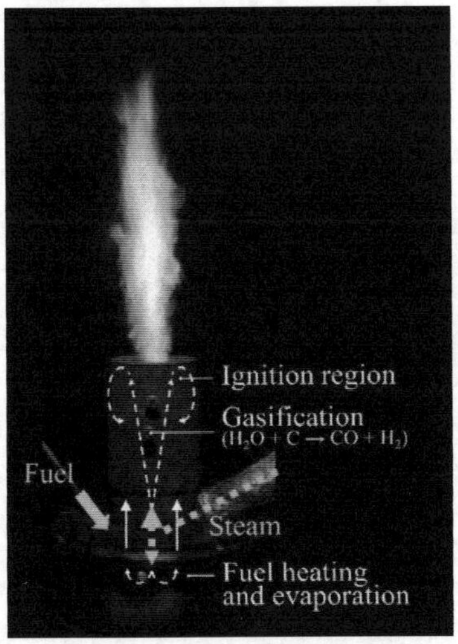

Figure 1. Flow chart of an evaporative burner operation with superheated steam forced supply.

Figure 2. Flow chart of spray burner operation

3. Comparative analysis of measurement results

To perform comparison, the regimes with the same thermal power (fuel flow rate Ff = 0.8 kg/h) were selected. The relative flow rate of steam (or heated air) γ, equal to the ratio of the mass flow rate of steam Fv (or heated air Fa) to the flow rate of the supplied fuel Ff, was varied. According to [16], the spray burner had a number of restrictions on the flow rate of the supplied steam (or air) associated with flameout at an oxidizer excess, as well as with underburning at its lack. Therefore, for the evaporative burner, the values of γ are given in the range from 0.3 to 1.5 for steam and from 0.7 to 1.5 for air, and for the spray burner, these values are given in the range from 0.3 to 0.8 for steam and from 0.5 to 1.0 for air at a power of 10 kW.

Using the data presented in [15,16], the dependences of contents of toxic components CO and NOx in the exhaust gases on the relative flow rate of superheated steam are constructed (Fig. 3-a, b). The values shown in the diagrams are recalculated from the volumetric values [ppm] (shown in the original works) to the mass content of a substance, referred to 1 kg of burnt fuel [g/kg] according to the method described in [17].

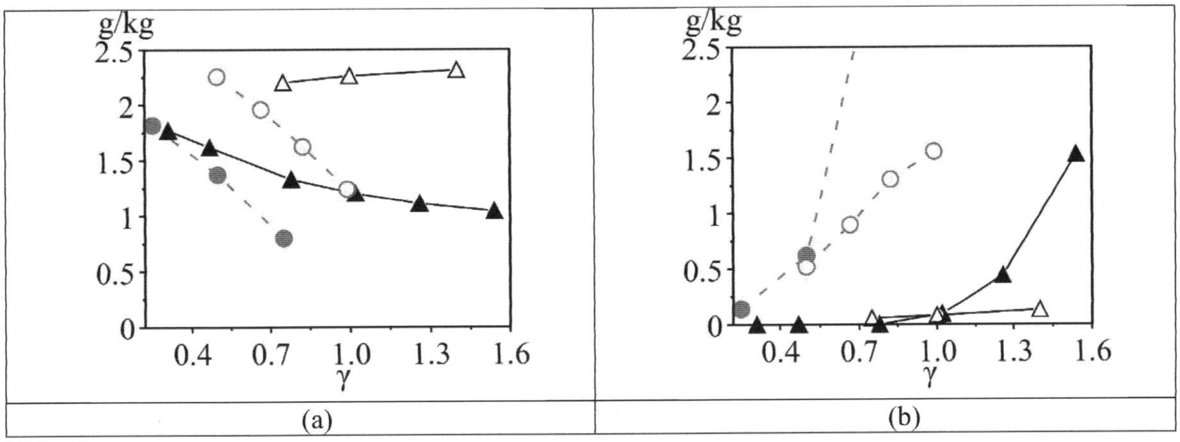

Figure 3. Contents of (a) nitrogen oxides, (b) carbon monoxide in exhaust gases for different relative flow rates of steam (solid symbols) or air (hollow symbols) during fuel combustion in evaporative (triangles) and spray (circles) burners at a power of 10 kW.

Figure 3-a shows that with the supply of superheated steam, a decrease in NOx content of about 30% is observed in comparison with the supply of heated air for both designs. However, in the spray burner, this increases the production of carbon monoxide (Fig. 3-b), probably associated with partial underburning of fuel when approaching the regime with the flameout, while for the evaporative burner the contents of CO remain minimal in a wide range of γ. For an evaporative burner, the optimal is the regime with $\gamma = 0.9$, for which the mass content of CO is within the measuring error of the device at [NOx] = 1.27 g/kg, which meets the 3 class of EN: 267 [17].

A high CO content in the spray burner may be due to the selected non-optimal power mode of 10 kW. Therefore, for a spray burner we take the parameters corresponding to the power of 15 kW (Ff = 1.2 kg/h) and compare them with the results obtained for an evaporative burner at the power of 10 kW (Ff = 0.8 kg/h). The obtained dependences of the [CO] and [NOx] content in the exhaust gases at different regimes of burner operation are shown in Fig. 4-a, b.

According to the figures, for all the presented regimes of spray burner operation, the content of carbon monoxide is minimal when either superheated steam or air is supplied at a device power of 15 kW. At that, the spray burner implements the regimes with lower NOx emissions as compared to the evaporative burner, in terms of 1 kg of burnt fuel. To explain this behavior, Fig. 5 shows distributions of the average temperature along the symmetry axis of the flame for the compared regimes of burner operation. It can be seen that when combusting diesel fuel in the evaporative burner, the higher average temperatures along the flame axis of symmetry are achieved. The average difference is about 250 degrees. This explains the different contents of nitrogen oxide for the investigated devices. Due to the higher temperatures more “thermal” nitrogen oxide forms. This behavior of average temperatures can lead to the fact that in the spray burner the jet of superheated steam or heated air, supplied instead, interacts directly with the fuel; due to this, the degree of gasification of fuel droplets can be increased. In addition, a second nozzle in the evaporative burner can reduce the degree of air entrainment from the environment, thereby reducing the “dilution” of the combustible mixture with cold air nitrogen, which also affects the temperature in the external flame of the device. Namely, when calculating the impulse of the supplied jet, which is directly related to the amount of ambient air sucked into the burner, according to the method described in [21], it can be estimated that the impulse of the supplied steam jet for a spray burner at $\gamma = 0.85$ and a power of 15 kW is 0.268 N, while for an evaporating burner at $\gamma = 0.8$ and a power of 10 kW it is 0.133 N. It can be seen that with a power difference of 1.5 times, the ratio of jet pulses is 2.

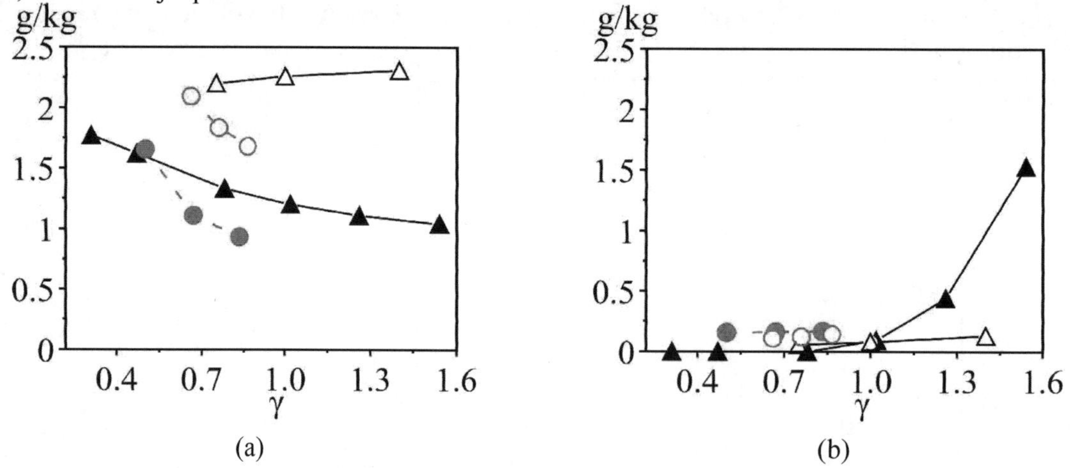

Figure 4. Content of (a) nitrogen oxides, (b) carbon monoxide in exhaust gases for different γ in the case of steam (solid symbols) or air (hollow symbols) supply during fuel combustion in evaporative (black triangles) and spray (red circles) burners at power of 10 kW and 15 kW, respectively.

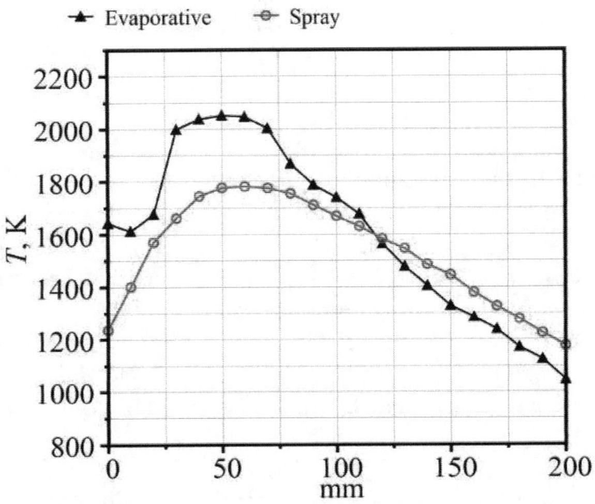

Figure 5. Distribution of the time-averaged temperature in the external flame along the axis of symmetry during fuel combustion with the supply of superheated steam at $\gamma = 0.8$ in evaporative (black triangles) and at $\gamma = 0.85$ spray (red circles) burners at power of 10 kW and 15 kW, respectively.

Thus, in spite of the lower indicators of the harmful substance content in exhaust gases for the spray burner, its operation with natural air entrainment from the environment is likely to require more air, which may be an obstacle to its use in areas with strict restrictions on this parameter. Therefore, it is difficult to draw a conclusion about the unambiguous advantage of one or another method of burning liquid fuel with the supply of superheated steam to the reaction zone. Both schemes under study, implemented in different burner designs, have their own disadvantages and advantages and can be successfully used to solve various problems.

Nevertheless, it is advisable to modify the design of the spray burner to organize controlled air supply, both inside the burner and for the ignition of reaction products in the external flame, and to experimentally determine the effect of the amount of incoming air on characteristics of device operation and optimization of operating parameters.

Conclusion

In this work, we compare and analyze previously obtained experimental data on the combustion of liquid hydrocarbon fuels in the presence of superheated steam in two different original designs of laboratory burners. These burners differ in the way of mixing superheated steam and fuel and feeding them into the zone of combustion.

It is shown that when combusting fuel in the spray burner, there are lower values of nitrogen oxide content in the exhaust gases than in the evaporative burner. However, when the burners operate on the same amount of fuel, at equal heat outputs, in the regime with low NOx emissions, a higher content of carbon monoxide in the exhaust gases is observed on the spray burner. Whereas an evaporative burner is characterized by a wide operating range, in which the CO values do not exceed the error of the measured device. In the operating regime, when the power of the spray burner is changed by 15 kW, a spray burner is shown to provide a lower emission of nitrogen oxides than an evaporative burner with simultaneously low emission of carbon monoxide. The low NOx content in the combustion products of the spray burner is associated with a lower average flame temperature due to the direct interaction of the steam jet with the fuel. The results obtained can be used to verify the mathematical model and numerical simulation of the process, as well as in the design of low-emission burners.

Acknowledgments
The study was financially supported by the Russian Science Foundation (project No. 18-79-10134, https://rscf.ru/project/18-79-10134/).

References

[1] Zhao N, Li B, Chen D, Ahmad R, Zhu Y, Li G, Yu Z, Li J, Wang E, Yun S, Yoon H, Yoon I, Zhou Y, Dong R, Wang H, Cao J, He J and Ju X 2020 *Waste Management* **104** 20
[2] Basler B and Felix P C 1982 in *Volume 5: Manufacturing Materials and Metallurgy; Ceramics; Structures and Dynamics; Controls, Diagnostics and Instrumentation; Education; Process Industries; Technology Resources; General*
[3] Göke S, Schimek S, Terhaar S, Reichel T, Göckeler K, Krüger O, Fleck J, Griebel P and Paschereit C O 2014 *Journal of Engineering for Gas Turbines and Power* **136** 091508
[4] Alekseenko S V., Pashchenko S É and Salomatov V V. 2010 *Journal of Engineering Physics and Thermophysics* **83** 729
[5] Li A, Zheng Z and Peng T 2020 *Fuel* **268** 117376
[6] Cui G, Dong Z, Wang S, Xing X, Shan T and Li Z 2020 *Applied Energy* **259** 114205
[7] Le Cong T and Dagaut P 2009 *Energy & Fuels* **23** 725
[8] Zou C, Song Y, Li G, Cao S, He Y and Zheng C 2014 *International Journal of Heat and Mass Transfer* **75** 12
[9] Albin E, Nawroth H, Göke S, D'Angelo Y and Paschereit C O 2013 *Fuel Processing Technology* **107** 27
[10] Boushaki T, Dhué Y, Selle L, Ferret B and Poinsot T 2012 *International Journal of Hydrogen Energy* **37** 9412
[11] Honzawa T, Kai R, Seino M, Nishiie T, Suzuki Y, Okada A, Wazaki K and Kurose R 2020 *Journal of Natural Gas Science and Engineering* **76** 103158
[12] Furuhata T, Kawata T, Mizukoshi N and Arai M 2010 *Fuel* **89** 3119
[13] Lellek S, Barfuß C and Sattelmayer T 2017 *Journal of Engineering for Gas Turbines and Power* **139** 021506
[14] Farokhipour A, Hamidpour E and Amani E 2018 *Fuel* **212** 173
[15] Anufriev I S, Alekseenko S V., Sharypov O V. and Kopyev E P 2019 *Fuel* **254** 115723
[16] Anufriev I S and Kopyev E P 2019 *Fuel Processing Technology* **192** 154
[17] DIN EN 267:2011-11. Automatic forced draught burners for liquid fuels.
[18] Anufriev I S, Alekseenko S V., Kopyev E P and Sharypov O V. 2019 *Journal of Engineering Thermophysics* **28** 324
[19] Kopyev E, Sharypov O, Anufriev I and Vigriyanov M 2020 *AIP Conference Proceedings* **2304** 020010
[20] Vigriyanov M S, Salomatov V V and Alekseenko S.V. 2003
[21] Anufriev I S, Krasinsky D V., Shadrin E Y, Kopyev E P and Sharypov O V 2019 *Thermophysics and Aeromechanics* **26** 657

Journal of Physics: Conference Series **2150** (2022) 012010 doi:10.1088/1742-6596/2150/1/012010

Problem of maintaining a normal condensate dissolved oxygen concentration

A D Vodeniktov and N D Chichirova

Kazan State Power Engineering University, str. Krasnoselskaya, 51, 420066, Kazan, Russia.

Vodeniktov@yahoo.com

Abstract. Study presents the results of the steam surface KCS-200-2 reconstruction. In order to maintain the dissolved oxygen concentration, according to the Code of Operation for Power Plants, the steam sparger was installed in condenser hotwell. Despite the abnormal air leakage level, reducing of dissolved oxygen concentration was reached. The dissolved oxygen concentration reduced, on average, by 2 times. As it was expected, in the cases of low inlet cooling water temperature, the final oxygen concentration did not reach the normal level. In last 3 tests the dissolved oxygen concentration was reduced to 19 mg /l. The results show a possibility of described reconstruction experience.

1. Introduction

The most important duty in the operation of steam turbines of thermal and nuclear power plants is to prevent corrosion of the entire feed-water circuit. Obviously, this is possible only if the water-chemical regime is maintained.

The presence of corrosive gases, such as oxygen and carbon dioxide in the main condensate of the turbine contributes to the intensification of corrosion of pipelines and equipment of the regenerative feedheating system in the condenser-deaerator section. Long work of the steam turbine during excess of dissolved oxygen concentration in feedwater systems has critical impact on the power plant reliability. The financial consequences of high dissolved oxygen concentration could be estimated as the increased repair costs. Figure 1 shows the dependence of the dissolved oxygen concentration on the rate of steel corrosion. It is linear; therefore, with an increase in dissolved oxygen concentration, the number of accidents caused by corrosion damage increases [1].

As the practice of operation shows, several factors have a significant impact on the quality of deaeration, which depend on the operating mode of the turbine and the circulation system.

Content from this work may be used under the terms of the Creative Commons Attribution 3.0 licence. Any further distribution of this work must maintain attribution to the author(s) and the title of the work, journal citation and DOI.
Published under licence by IOP Publishing Ltd

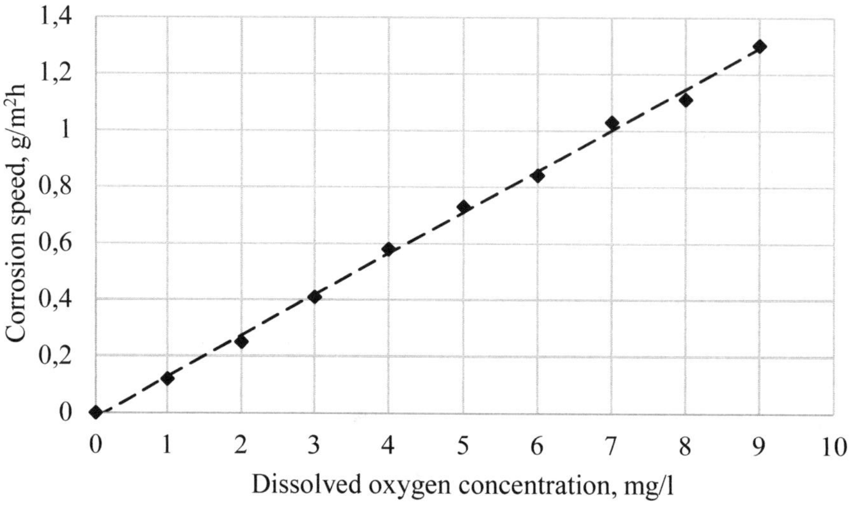

Figure 1. Dependence of the corrosion rate on dissolved oxygen concentration

2. Problem description

According to [2], both condensing and heating turbines face a problem of excess concentrations of dissolved oxygen measured at the outlet of condensate pumps relative to the values established by the rules of technical operation of electric stations and networks [3].

One of the main factors determining oxygen concentration in the turbine condensate is the steam load of the condenser (condenser duty), which depends on the operating mode of the steam turbine. According to [4], deep deaeration of the exhaust steam condensate in the turbine condenser is possible in a wide range of loads. However, when the steam loads of the condenser are below 60-40% of the nominal, there is deterioration in the deaeration regime and an increase in the concentration of dissolved oxygen. This can be explained by the fact that with a decrease in the steam load of the condenser, the zone of mass condensation of steam shifts closer to the exhaust hood, and the cooling zone of the steam-air mixture begins to grow rapidly and go beyond the area of the air cooler. This problem is especially acute for heating turbines that operate for a long time according to a thermal schedule with a minimal steam flow into the condenser.

Separately, it is worth noting the influence of the air leakages on the oxygen content in the main condensate. All air leakages into the steam space of the condenser do not affect the deaeration regime in the condenser, since they are under the influence of the ejector and all the air is removed by the ejector (during normal operation of the ejector) [4]. On the other hand, the closer the air leakage is to the hotwell, the greater the effect it has on aeration of the condensate flowing down from the tubes. The critical moment is the presence of air-leakage in the water volume of the condenser. In this case, all the incoming air cannot be removed by the ejector and is completely mixed with the condensate, saturating it.

Also, the deaerating ability of the condenser is affected by the temperature of the cooling water. It was noted in [4] that for regenerative condensers, the deaeration ability does not decrease even with a decrease in the temperature of the cooling water. However, [5] shows the negative effect of the cooling water temperature lowering on the water-chemical regime of the condensers of the modern steam turbines. The deterioration of deaeration was observed when the temperature of the cooling water at the inlet to the condenser was below 11°C.

Dissolved oxygen produced by sub-cooling in the condenser tube bundle, not heated by steam entering from the bottom of the bundle, will enter the hotwell and will be diluted by condensate, which is not sub-cooled and aerated. Once this saturated with oxygen condensate falls below the bottom of the tube bundle and comes through the liquid surface of the hotwell, the dissolved gases most likely will

not be released primarily due to the head pressure of water from above. Therefore, it contributes to the increasing of dissolved oxygen concentration in the condensate.
The above analysis indicates that it is impossible to provide the necessary concentration of dissolved oxygen in the main condensate in the condenser-deaerator section for steam turbines. There remains the necessity of special additional measures to intensify desorption of dissolved oxygen in the steam turbine condenser.

3. Actual situation

The process of vapor condensation in the condenser tube bundle is accompanied by mass transfer between non-condensable gases (air) in the vapor and the liquid phase (condensate).
Experience shows that in regenerative condensers, the condensate at the outlet has low concentrations of dissolved oxygen. Thus, the condenser is the first deaeration stage.
Inside the condensers, it is possible to deaerate the condensate directly on the surface of the tubes. In this case, the final concentration of dissolved oxygen depends on the concentration of air in the steam, the steam flow rate, the flow paths of steam and water, and the temperature of cooling water and condensate.

However, to ensure the required quality of deaeration, especially at start-up modes, it is necessary to install a special device.
Various models of steam turbine condenser have installed deaerating devices inside the hotwell. In the designs of condensers for the Soviet and foreign turbines, this solution has been used for more than 60 years. Depending on the flow path of the condensate and the heating fluid, deaeration devices of the condensers are classified as follows: film-type, nozzle-type, jet-type (when water moves in steam), and bubbling-type (when steam moves in water).

Similar devices are used on K-500-60/1500 HTGZ (Fig. 2), K-300-240 LMZ and T-175/210-130 UTZ, and many others turbines. As the thermal tests have shown, these devices allow maintaining the required concentration of the dissolved oxygen in the main condensate in the most operating regimes of the steam turbine.

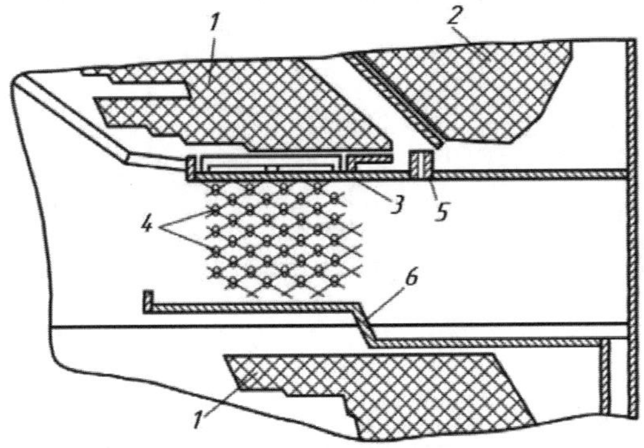

Figure 2. Deaerating device of the K-22550 HTGZ surface condenser. 1-tube bundle, 2-air cooler, 3-water distributing sheet, 4- uncooled tubes, 5-non-condensable vent, 6-water collecting sheet.

However, some of the most numerous turbines, such as K-210-130, do not have special deaeration devices inside the condenser. Together with a long service life and an unsatisfactory condition of the vacuum system, operation regimes with increased dissolved oxygen concentration are often observed.

4. Case description

The authors, in accordance with the recommendations of [6], have reconstructed and studied the operating modes of the 200-KCS-2 steam surface condenser. The reconstruction included the installation of a "steam sparger" type device inside the hotwell. The principle of its work was condensate heating to the saturation temperature.

The condenser tube bundle scheme is presented in Fig. 3. This design is typical of the regenerative condensers of 60-70 years. The arrangement of tubes is the tape-like. The condenser is made of two-path, with a side arrangement of the air cooler.

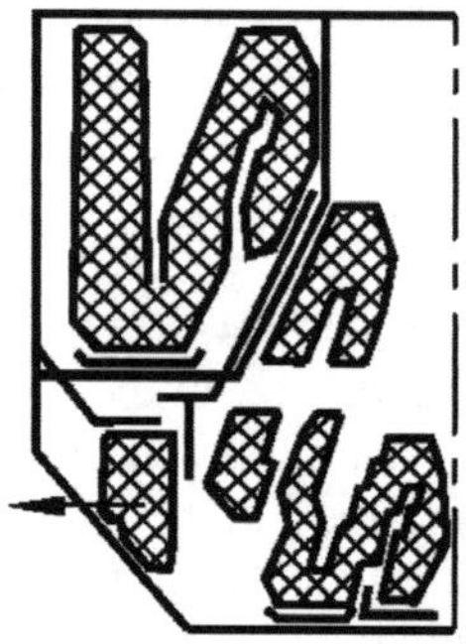

Figure 3. Tube bundle scheme of 200-KCS-2 condenser.

Thermal tests were carried out with the nominal load of the steam turbine at different temperatures of cooling water at the condenser inlet. The lower temperatures of cooling water matched a higher concentration of dissolved oxygen. Throughout the tests, the air leakage rate of the vacuum system remained constant and exceeded the norm. The values of the dissolved oxygen concentration in the condensate were recorded after the steam-sparger was switched on after 15 minutes, to stabilize the water-chemical regime. Stationary chemical control devices were used as the oxygen meter.

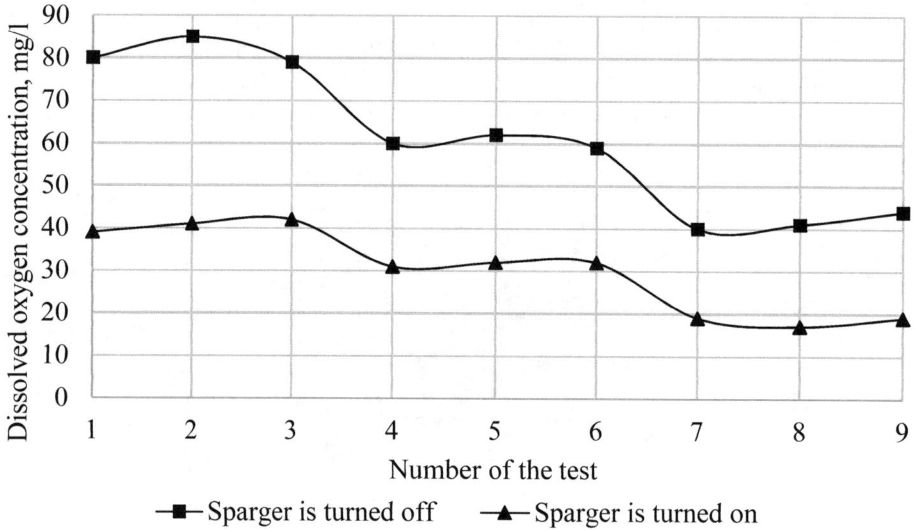

Fig. 4. Dissolved oxygen concentration reducing during thermal tests.

The use of superheated steam from the regeneration system allowed necessary heating of the condensate. During the tests of the turbine unit at a load range of 100-40%, the activation of the device in operation allowed reduction in oxygen concentration in the main condensate by 2-2.5 times to the norms established by the PTE.

5. Conclusion

Based on the obtained results, the use of deaeration devices in turbine condensers seems appropriate to maintain the water-chemical regime over the entire range of loads.

As it is shown by the test results, the installation of a steam sparger increases the deaeration capacity. However, additional use of steam increases the specific heat consumption.

Undoubtedly, the high value of air leakages in the vacuum system, especially in the water volume of the condenser, significantly worsens the deaeration of water in the condenser.

It should be noted that in some cases, a decrease in the concentration of dissolved oxygen did not allow reaching the regulated values. To improve the efficiency of the steam sparger, both a design change and an increase in steam flow rate are necessary.

References

[1] Akolzin P A, Corrosion of structural materials of nuclear and thermal power plants. High School. Moscow, 1963, 376p.
[2] Shempelev AG, Iglin P.V., The reasons of increased oxygen content in the condensate in the condenser during operation close to the nominal operating conditions. Society. Science. Innovations (NPK-2017). Kirov: Vyatka State University, 2017. p. 2385-2391.
[3] Russian Federation Code of Operation for Electrical Power Plants and Grids
[4] Kirsh A.K. Deaeration in steam surface condensers. Moscow, BTI ORGRES, 1960, 28 p.
[5] Vodeniktov A D, Chichirova N D. Influence of the temperature of the cooling water on the deaeration capacity in the KCS-200-2 condenser. Transactions of Academenergo. 2020;4 (61)
[6] Standards for steam surface condensers. 12th edition. HEI, 2017, 116 p.

Influence of the initial parameters on the thermodynamic efficiency of carbon dioxide power cycles

P A Shchinnikov[1], I S Sadkin[1,2], A P Shchinnikov[3], N F Cheganova[1] and N I Vorogushina[1]

[1]Novosibirsk State Technical University, Novosibirsk, Russia
[2]Kutateladze Institute of Thermophysics SB RAS, Novosibirsk, Russia
[3]Moscow Power Engineering Institute, Moscow, Russia

sadkinvanya@mail.ru

Abstract. This paper considers the main CO_2 power cycle configurations based on the Allam and JIHT cycles. In particular, the authors of the article have proposed new configurations of the power cycle. The efficiency of these cycles is studied as a function of the initial temperature and pressure of the working fluid. The thermodynamic efficiency can reach 65–66%. It is shown that the presence of regenerative heat transfer and the properties of supercritical carbon dioxide have a great influence on the thermal efficiency.

Introduction

At present, technologies decreasing the carbon footprint are becoming more and more relevant. The reason for this is the efforts of the world community to reduce anthropogenic greenhouse gas emissions in furtherance of sustainable development and global warming mitigation [1–3].

Therefore, nonconventional and renewable power sources that reduce the consumption of hydrocarbon fuels have been actively developed in power engineering. However, it is equally important to develop technical solutions in the field of using traditional fossil fuels that can reduce or completely eliminate the carbon footprint. These technical solutions include carbon dioxide based power cycles, in which the working fluid of the cycle is supercritical carbon dioxide. Currently, the best-known and widely described in the literaturecycles are the JIHT cycle [4] and the Allam cycle [5], which use almost pure carbon dioxide as a working fluid, and differ from each other in that condensation of the working fluid occurs in the JIHT cycle and does not occur in the Allam cycle. In addition, there is the theoretical Graz cycle [6], which uses a mixture of carbon dioxide and water vapor as a working fluid. In all cases, CO_2 is utilized by liquefaction and disposal. Gas is used as a fuel, and oxygen as an oxidizer; therefore, the combustion products are only CO_2 and water vapor resulting from combustion of fuel hydrogen. After the expansion of the working fluid in the turbine, water condenses and pure carbon dioxide remains, which can be recycled in another process or disposed. Therefore, the study of carbon dioxide power cycles is an urgent task.

Research object and method

In this paper, the main carbon dioxide power cycle configurations are considered (see Figure 1).

Content from this work may be used under the terms of the Creative Commons Attribution 3.0 licence. Any further distribution of this work must maintain attribution to the author(s) and the title of the work, journal citation and DOI.
Published under licence by IOP Publishing Ltd

Cycle 1 is a simplified variant of the JIHT cycle, and cycle 4 is a simplified variant of the Allam cycle. Cycle configurations 2 and 3 are proposed by the authors of the paper. These cycle configurations and features of their operation are described in more detail in [7].

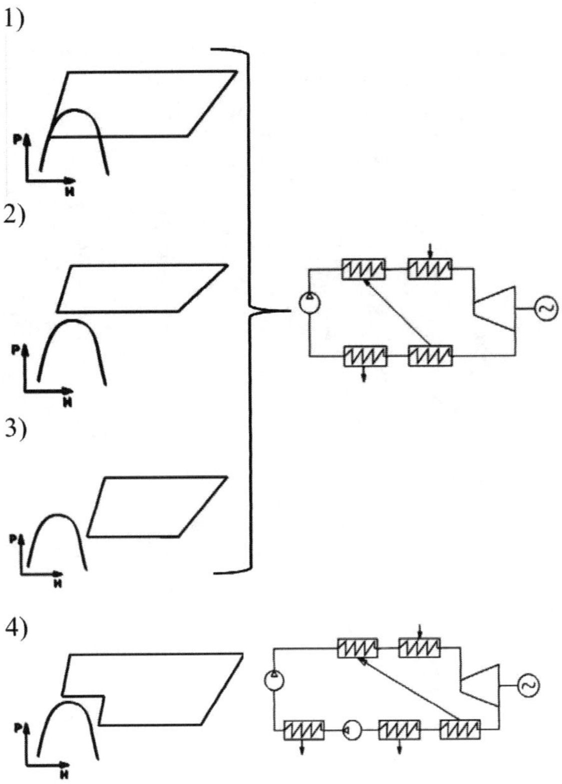

Figure 1. Thermodynamic CO_2 cycles and schematic diagrams of power generation

The efficiency of these cycle configurations (Figure 1) was analyzed in [7]. The calculation showed that the thermal efficiency of all cycles exceeds that of the traditional single-stage cycles used today in the power industry. The efficiency of the presented cycle variants is 57–63% at the initial parameters (temperature of 1300 °C, pressure of 300 bar) similar to those declared in the Allam cycle [5]. The final pressures of the considered cycle variants differ from each other, Table 1. This is explained by the technological features of each cycle [7].

Table 1. Final pressure of the considered cycles

Cycle number	1	2	3	4
Final pressure, bar	57,2	75	30	30

It is also important that the thermal efficiency of the cycles proposed by the authors of the article is not only comparable to but in some cases even higher than that of the Allam and JIHT cycles widely described in the literature. However, these variants of cycles have not been reported in the literature, so it is advisable to study them further.

Here we analyze the thermodynamic efficiency of the cycles depending on the initial pressure over a wide temperature range. At this stage, the efficiency of possible equipment that can be used at a real power plant, costs for own needs (for example, for fuel supply) are not considered. Only costs of compression (in isentropic process) of CO_2 removed from the cycle are taken into account. The calculation method used for this is based on the material and heat balance equations for the combustion chamber and the regenerative system described in [7] and thermodynamic tables and phase diagrams for CO_2 and water vapor. Methane is selected as fuel for calculation in all cycle configurations.

Results and discussions

The thermal efficiency of the cycles is calculated in the temperature range 1000–1500 °C for three initial pressures: 200, 300, and 400 bar. The calculation results are presented graphically in Figure 2. The numbers on the curves corresponds to the type of cycle in Figure 1.

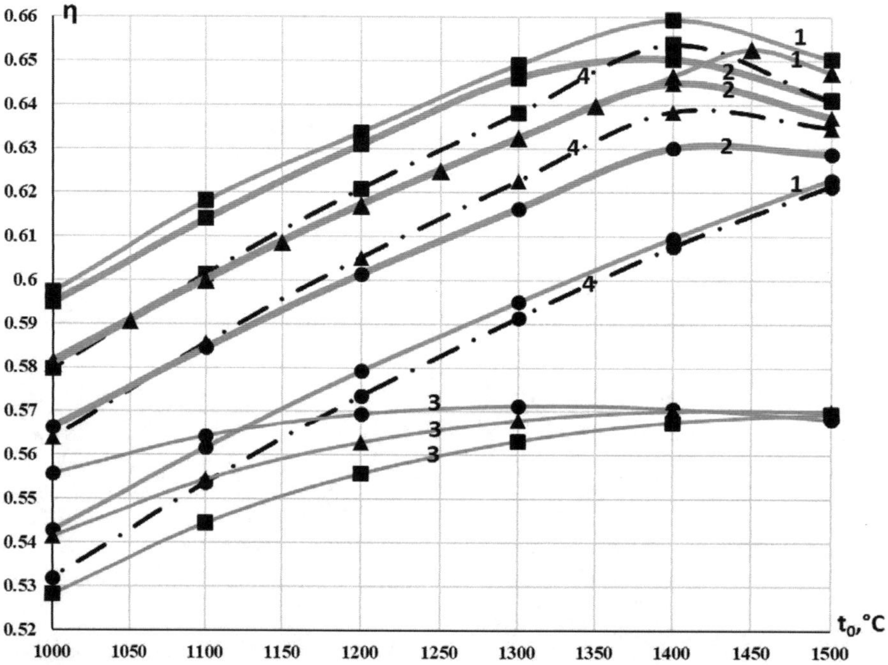

Figure 2. Curves of the thermal efficiency of cycles versus initial temperature at different initial pressures of the working fluid. Legend: curves with squares correspond to an initial pressure of 400 bar, curves with triangles to 300 bar, and curves with circles to 200 bar. The number on the line indicates the type of cycle.

At an initial pressure of 200 and 300 bar over almost the entire temperature range, cycle 2, entirely located in the zone of supercritical parameters and proposed by the authors of the paper, has the greatest efficiency. Cycle 1 (simplified JIHT, with condensation) ranks second in efficiency, and cycle 4 (simplified Allam, without phase transition) ranks third. Cycle 3 shows the least efficiency. When the pressure rises to 400 bar, the cycle 1 (JIHT) has the greatest efficiency, cycle 2 (proposed by the authors) ranks the second.

It can be seen that as the initial pressure increases from 200 to 400 bar in the investigated temperature range 1000–1500 °C, the efficiency of cycles 1, 2, and 4 increases, and the thermal efficiency of cycle 3 decreases. This is due to the fact that cycle 3 involves only compression of carbon dioxide in the compressor to the operating pressure of the cycle. This variant is much more energy intensive and leads to the fact that the increase in the heat drop due to an increase in the initial pressure is slower than the increase in the cost of compressing the working fluid in the compressor. As the pressure increases from 200 to 400 bar, the available heat drop increases by a factor of 1.33, and the compression power consumption increases by a factor of 1.54.

In cycles 1, 2, and 4, the pressure rises in two stages: first in the compressor to 75 bar and then in the pump to the operating pressure. Therefore, in these cycle configurations, there is no sharp increase in power consumption for compression as the initial pressure increases.

At the same time, cycles 1, 2, and 4 are characterized by an inflection of the efficiency curves. This is due to the influence of the heat regenerator and the properties of carbon dioxide. This happens as follows. An increase in the initial temperature of the working fluid leads to the need to burn more fuel, which results in an increase in combustion products and an increase in the fraction δ of CO_2 introduced into the cycle from fuel combustion. Then exactly this amount of carbon dioxide is removed from the cycle and disposed of after expansion in the turbine and passage through the regenerative heat exchanger. The CO_2 flow reduced by the value of δ returns to the combustion chamber and is supplemented with combustion products, and the cycle repeats. However, before entering the combustion chamber, this flow is heated in the regenerative heat exchanger by the carbon dioxide flow coming out of the turbine, which exceeds the heated flow by the value of δ and the mass of water vapor formed from the oxidation of fuel hydrogen. At the same time, the heated CO_2 stream is under the operating pressure of the cycle (about 10 times higher than the turbine exhaust pressure) and has a higher heat capacity due to the properties of CO_2. Usually, the difference in heat capacity allows the heated flow to receive all heat of the larger heating flow of CO_2 and water vapor. However, at a certain moment, the quantity 'δ' reaches such a value that the heated CO_2 stream is no longer able to accept the entire amount of heat of the regenerator, and part of the heat is forced into the atmosphere and does not remain in the cycle, which leads to a decrease in the overall thermodynamic efficiency and the appearance of an inflection in the curve.

As the initial pressure of the cycle increases, the heat capacity of carbon dioxide also increases. This leads to the need to burn even more fuel to achieve a high temperature. Because of this, the value of 'δ increases over the entire range of initial temperatures, which leads to the fact that even at lower initial temperatures of CO_2, a situation arises where not all the heat of the working fluid coming out of the turbine is used by the regenerator. The inflection of the curve is shifted to the left.

For the same reason, the cycle efficiency curves for different initial pressures stick together as the initial temperature increases.

Conclusions

The main configurations of single-stage carbon dioxide power cycles were considered, and their thermodynamic efficiency was determined at various initial pressures (200, 300, 400 bar) in a wide range of initial temperatures (1000–1500°C). The thermodynamic efficiency can reach 65–66%, which is consistent with literature data on the well-known similar cycles (Allam cycle and JIHT cycle). It is shown that the presence of regenerative heat transfer and the properties of supercritical carbon dioxide have a great influence on the dependence of the thermal efficiency on the initial parameters, which requires a more detailed study when considering more complex cycle configurations.

References

[1] Meng-Tian Huang, Pan-Mao Zhai 2021 *Advances in Climate Change Research* **12(2)** 281-286
[2] D.M. Smith, A.A. Scaife, E. Hawkins et al. 2018 *Geophysical Research Letters*, **45(21)** 11895-11903
[3] Leung D Y C., Caramanna G, Maroto-Valer M M 2014 *Renew. Sustain. Energy Rev* **36** 426-443
[4] Kosoi A S, Zeigarnik Yu A, Popel' O S, Sinkevich M V, Shterenberg V Ya 2018 *Thermal Engineering* **65(9)** 597-605
[5] Allam R, Martin S and others 2017 *Energy Procedia* **114** 5948-5966
[6] Wolfgang Sanz Graz Cycle. – A Zero Emission Power Plant for CCS (Carbon Capture and Storage) http://www.graz-cycle.tugraz.at/
[7] Shchinnikov P, Borush O, Frantseva A, Sadkin I 2021 *E3S Web of Conferences* **289** 02001

Intensification of heat and mass transfer of the gas channel of the multi-temperature condensation filter

M V Malevany and D A Konovalov

Voronezh State Technical University "VSTU",
Russia, 394006 Voronezh, 20-letiya Oktybrya, 84

dmikonovalov@yandex.ru

Abstract. The principle of operation and features of hydrodynamics and heat-mass transfer in the working channel of a multi-temperature condensation filter for gas purification are described. Promising methods of gas flow purification using porous surfaces are described. The modeling and comparison of the laminar air flow in channels of various shapes: in a flat channel, in a channel with webs and in a spiral channel are performed. An analysis of their effectiveness is carried out.

1. Introduction

The issue of purifying gas streams from various mechanical and chemical impurities is of particular interest, since, in addition to purifying gas to obtain a purified stream as an end product, it is often necessary to capture valuable suspended particles contained in the purified gases. These requirements for the purification of streams can be implemented with a sufficient degree of efficiency in a multi-temperature condensation filter [1], the working channel of which is formed by walls having different temperatures (figure 1).

Figure 1. Diagram of a working channel for a multi-temperature condensation filter: a – cold wall; b – hot wall; c – gas input; d – gas output

The gas stream is purified by supplying a contaminated gas stream to a working channel at different temperatures. During its passage a zone of stable supersaturation is formed in the purified stream, which promotes the formation and growth of liquid droplets on various inclusions contained in the stream. Part of the moisture with trapped inclusions is deposited on the colder wall of the working channel and is discharged outside. The rest is captured behind the working channel quite simply, the channel moisture droplets on suspended particles significantly coarsen.

Content from this work may be used under the terms of the Creative Commons Attribution 3.0 licence. Any further distribution of this work must maintain attribution to the author(s) and the title of the work, journal citation and DOI.
Published under licence by IOP Publishing Ltd

Currently, an increasingly promising direction in the purification of gas streams is the use of various developed surfaces with sorbing properties. So, in works [2-4], the processes of creation, application and development prospects of various polymer membranes for gas purification are described. The works [5-7] describe the use of various organometallic frameworks for gas separation and adsorption. Let us look at some of these examples below.

The work [8] describes the application of various membranes in industry, and shows examples of modifications of membrane surfaces that improve the reaction processes on these surfaces (figure 2).

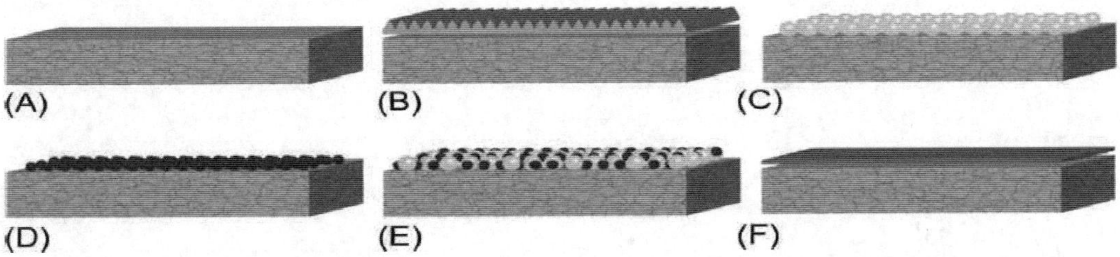

Figure 2. Schematic representation of some examples of surface modification membranes: (A) bare membrane, (B) etched membrane, (C) porous coated membrane, (D) catalyst particle deposition, (E) catalyst particle activated porous coating, and (F) protective layer.

Quite a lot of attention has been paid to work investigating CO_2 scrubbing on developed surfaces. In [9], various nitrogen-doped microporous coals were shown to demonstrate a high degree of CO_2 absorption (up to 197 mg / g) at 273 K and a pressure of 1 bar. The effective pore size was also determined to be 0.84 nm giving the best CO_2 uptake results. Work [10] described the creation of a cost-effective adsorbent in the form of porous anthracite carbons, created by a combination of KOH activation and treatment with urea to capture CO_2 and separate CO_2 / CH_4. The work [11] described the use of spongy pulp of a sunflower stem for the production of highly efficient porous carbon with the prospect of using it as a sorbent for CO_2, radioactive iodine, and oily / organic water pollution.

In devices such as a multi-temperature condensation filter (figure 1), a significant role is played by the efficiency of heat and mass transfer between the gas flow and the walls of the working channel [12, 13] with the purified gas flow. Thus, it becomes necessary to solve the problem of intensifying heat and mass transfer processes in the working channel.

The heat and mass transfer may be enhansed by using various developed surfaces [14-17], showing high efficiency due to the improvement of hydrodynamic and heat exchange indicators at the flow of liquids and air flows in the channels.

The purpose of this work is to assess the influence of developed surfaces and channel configuration on hydrodynamics and heat transfer in the channel.

2. Problem statement

One of the clearest ways to obtain a picture of the distribution of velocities, temperatures, etc., used in various studies [18-21], is the modeling of processes in software packages. In this work, we will use the COMSOL Multiphysics package to simulate a stationary gas heating process in a laminar gas flow through channels of various shapes.

It is required to make an approximate simulation of the gas heating process in laminar flow in channels of various shapes (figure 3) with the following similar data:
- three steel ducts of various shapes (flat duct, webed duct, spiral duct);
- conditionally constant rectangular channel section - 5mm x 20mm and channel length - 120mm;
- constant temperature of the channel walls - 80 ° C;
- working environment - air;
- air temperature at the inlet - 20 ° C;
- air velocity at the channel inlet – 1 m/s.

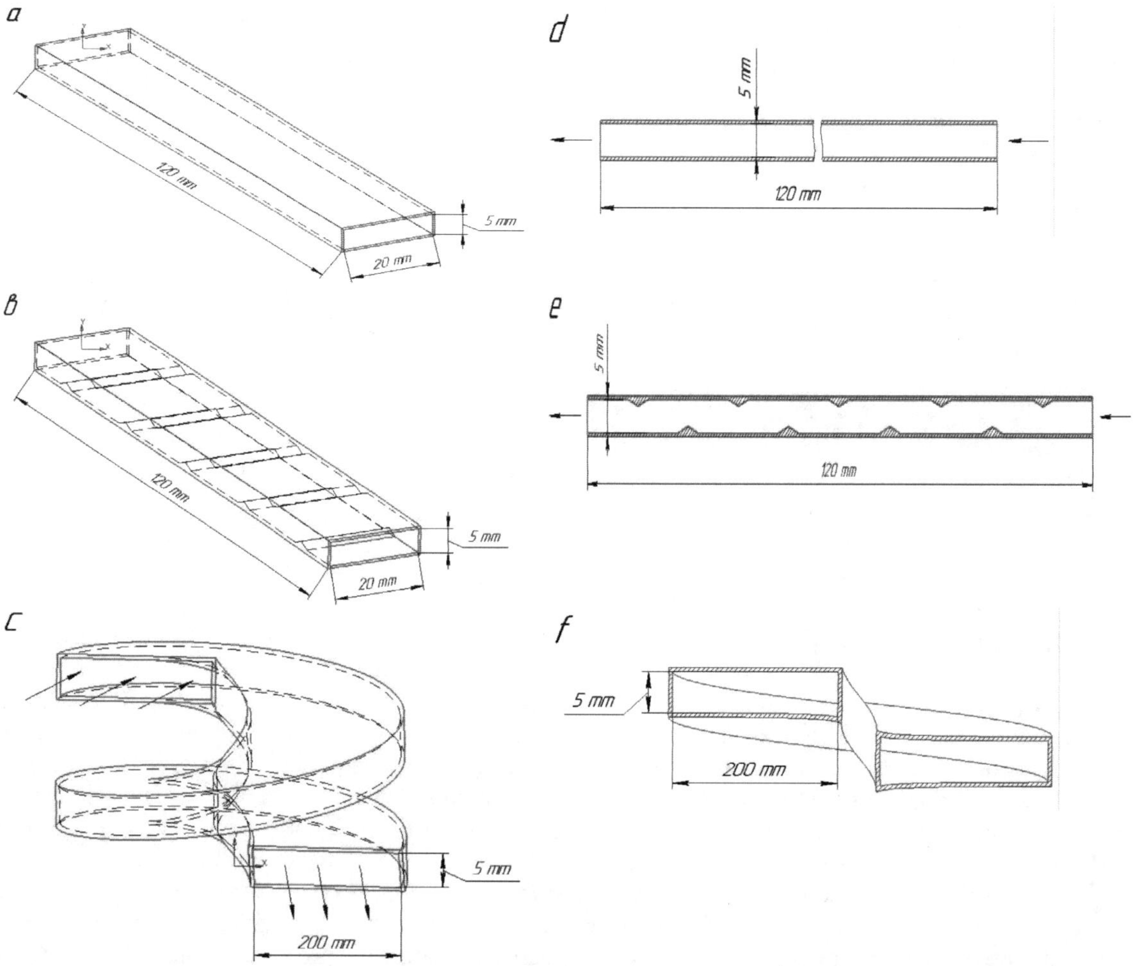

Figure 3. Channels of various shapes: flat channel (a), channel with projections (b), spiral channel (c); cross-sections: flat duct (d), lug duct (e), spiral duct (f)

When constructing a mathematical model, a number of assumptions were made: the flow is three-dimensional stationary; there is no initial hydrodynamic section at the channel entrance; there is no heat exchange with the environment; the thermophysical properties of the flow are taken to be constant and equal to the average values in the investigated temperature range.

The modeling used the interfaces "Heat Transfer in Solids and Fluids" and "Laminar Flow" with built-in solvers.

3. Results and discussion

As a result of the simulation, the values of temperatures and velocities were obtained in different regions for the three considered channels: for a flat channel (a, d), a channel with protrusions (b, e), a spiral channel (c, f) (figure 4).

On a flat channel: a gradual uniform heating of the gas is observed. The maximum temperature is observed in the near-wall regions, in the corners and at the exit from the channel, which corresponds to the laminar flow regime. The average temperature at the outlet from the channel is 69.1 ° C (figure 4, a). The maximum gas flow velocity is reached in the central region of the channel, while in the near-wall regions the velocity is insignificant and increases slightly closer to the exit from the channel (figure 4, d). The pressure drop across the channel is approximately 2.1 Pa.

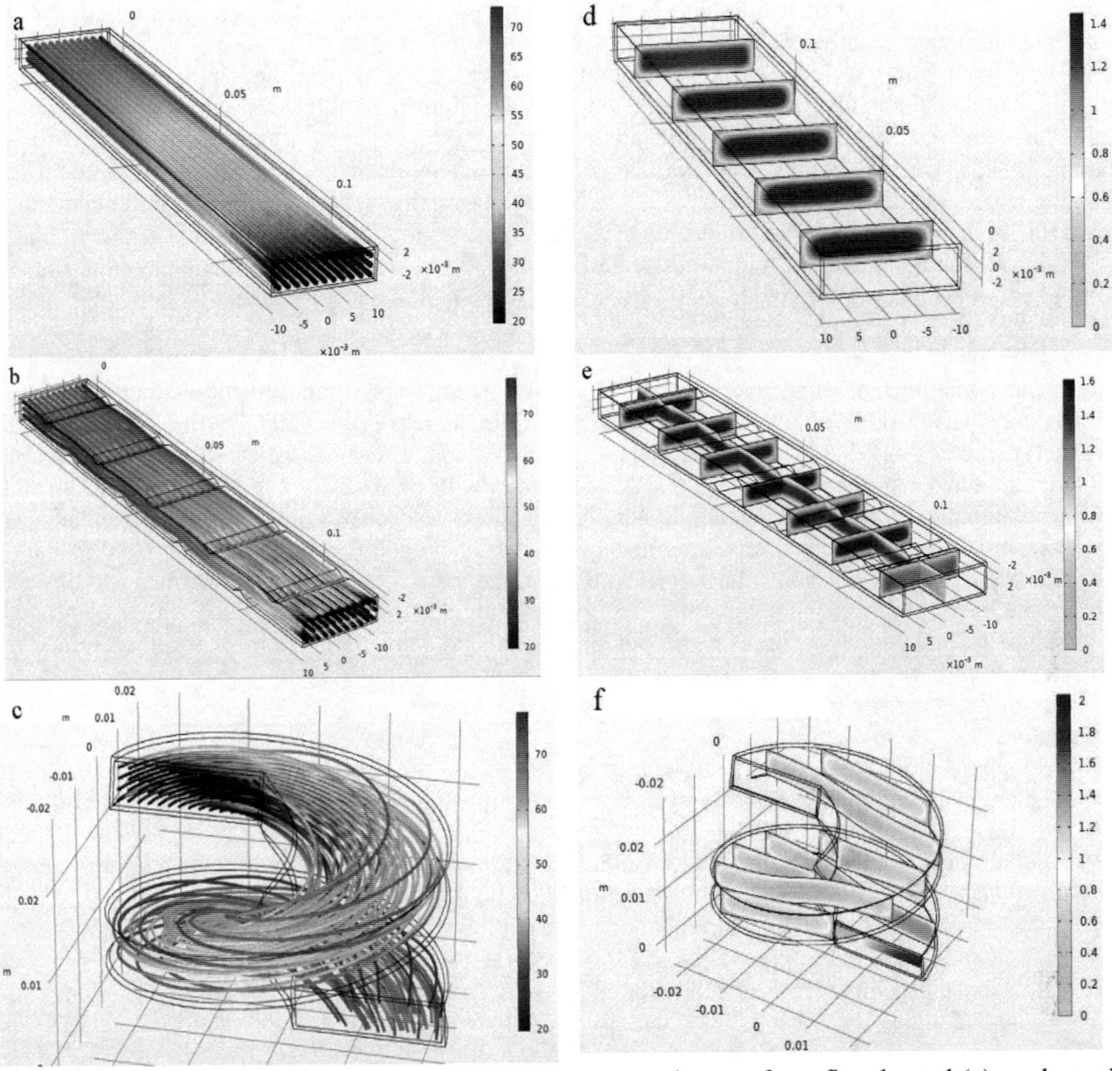

Figure 4. Simulation results: flow lines with temperature changes for a flat channel (a), a channel with protrusions (b), a spiral channel (c); cross-sections for visualizing flow velocities in a flat channel (d), a channel with protrusions (e), a spiral channel (f)

Through the channel with protrusions: a sharper heating of the gas is observed than in the flat channel. Thermal stabilization of the flow occurs in areas close to the channel entrance. The maximum temperature is observed in the same areas as in the flat channel, and near the protrusions as well. The average temperature at the outlet from the channel is 70.5 ° C (figure 4, b). The gas flow velocity pattern is similar to that of a flat channel, but there are increased velocities on the ridges and stagnant zones with low velocities in the areas between the ridges (figure 4, e), which is expected when a laminar flow moves in channels of a similar shape. The pressure drop across this lugged passage is approximately 3.0 Pa.

Along the spiral channel: significant heating of the gas is observed after passing the middle of the calculated region of motion. At the same time, in the near-wall regions along the inertia of the gas movement, the temperature is significantly higher than in other zones. Abundant mixing together with flow heating occurs after passing through the middle of the calculated region of motion. Average temperature at the outlet from the channel is 72.4 ° C (figure 4, c). The picture of the current velocity is very different from the two previously considered channels. In a spiral canal, the speed is fairly uniform and is approximately equal to the speed at the entrance to the canal (1 m / s) (figure 4, f). This can be explained by the constant mixing of the flow while moving in a spiral. The pressure drop across the spiral duct is approximately 3.6 Pa.

Even with a small scale of the computational domain (height, width and length of the channel 5x20x120 mm) and an air velocity of 1 m / s, we observe a significant improvement in thermohydraulic characteristics when using a channel with protrusions (figure 3, 4, b, e), as well as a spiral channel (figure 3, 4, c, f). The use of these channels shows an increase in the average outlet temperature by 1.5 ° C in the channel with protrusions and by 3.5 ° C in the spiral channel compared to the flat channel. In this case, an increase in the pressure drop by 0.9 Pa is observed in the channel with protrusions and by 1.5 Pa in the spiral channel as compared to the flat channel. Also in the spiral channel, a significantly homogeneous distribution of velocities is observed throughout the entire channel.

4. Conclusions

Under the conditions of our problem, we observed a significant improvement in thermohydraulic performance when using a channel with protrusions (figure 4, 4, b, e), as well as a spiral channel (figure 3, 4, c, f) compared to a flat channel (figure 3, 4 , a, d). The use of these channels showed an increase in the average outlet temperature by 1.5 ° C in the channel with protrusions and by 3.5 ° C in the spiral channel compared to the flat channel. In this case, an increase in the pressure drop by 0.9 Pa was observed in the channel with protrusions and by 1.5 Pa in the spiral channel as compared to the flat channel. Also, in the spiral channel, a significantly homogeneous distribution of velocities was observed throughout the channel.

The use of developed surfaces in the purification of gas streams is a promising direction in the development of industry.

References

[1] Ryazhskikh V I, Voronov G G, Solzhenikin P A. and Stogney V G 2014 Improvement of the engineering methodology for calculating the multi-temperature condensation filter *Bulletin of VSTU* Vol 10 No1 pp 107-9

[2] Endre Nagy 2019 Chapter 18 - Membrane Gas Separation *Basic Equations of Mass Transport Through a Membrane Layer (Second Edition)* pp 457-81

[3] Eric Favre 2017 2.6 Polymeric Membranes for Gas Separation *Comprehensive Membrane Science and Engineering (Second Edition)* vol 2 pp 124-75

[4] Kamran Ghasemzadeh, SadatiTilebon S M and AngeloBasile 2020 Chapter 5 - Membrane technologies for exhaust gas cleaning and carbon capture and sequestration *Current Trends and Future Developments on (Bio-) Membranes Cogeneration Systems and Membrane Technology* pp 97-123

[5] Hao Li, Libo Li, Rui-Biao Lin, Wei Zhou, Zhangjing Zhang, Shengchang Xiang and Banglin Chen 2019 Porous metal-organic frameworks for gas storage and separation: Status and challenges *EnergyChem* vol 1 Issue 1 100006

[6] Daniel Manaye Kabtamua, Yi-nan Wua and Fengting Li 2020 Intensification Hierarchically porous metal-organic frameworks: synthesis strategies, structure(s), and emerging applications in decontamination *Journal of Hazardous Materials* vol 397 122765

[7] Pelin Kinik F, Stavroula Kampouri, Fatmah Mish Ebrahim, Bardiya Valizadeh and Kyriakos Stylianou C 2021 2.30 - Porous Metal-Organic Frameworks for Advanced Applications *Comprehensive Coordination Chemistry III (Third Edition)* pp 590-616

[8] Julio Garcia-Fayos, José M. Serra, Mieke W.J. Luiten-Olieman and Wilhelm A 2020 8 - Gas separation ceramic membranes *Advanced Ceramics for Energy Conversion and Storage Elsevier Series on Advanced Ceramic Materials* pp 321-85

[9] Adeela Rehman and Soo-Jin Park 2019 Environmental remediation by microporous carbon: An efficient contender for CO_2 and methylene blue adsorption *Journal of CO_2 Utilization* vol 34 pp 656-67

[10] Yao Li, Nan Liu, Tao Zhang, Binbin Wang, Yan Wang, Lanyun Wang and Jianping We 2020 Highly microporous nitrogen-doped carbons from anthracite for effective CO_2 capture and CO_2/CH_4 separation *Energy* vol 211 118561

[11] Xin Li, Huan Zhang, Junfu Wei, Zhiyun Kong, Yanchen Zhao, Yuhong Liu, Changhao Li, Bowen Yang and Zhi Qiao 2020 Adsorptive nonwoven with multi-3-dimensional structure for the adsorption of VOCs *Microporous and Mesoporous Materials* vol 303 110190

[12] Ting Wang, En Lin, Yun-Lei Peng, Yao Chen, Peng Cheng and Zhenjie Zhang 2020 Rational design and synthesis of ultramicroporous metal-organic frameworks for gas separation *Coordination Chemistry Reviews* vol 423 213485

[13] Ryazhskih V I, Keller A V, Ryazhskih A V, Nikolenko A V and Dakhin S V 2020 Mathematical model of the acceleration laminar flow of a Newtonian fluid in an anisotropic porous channel of rectangular cross section *Bulletin of the South Ural State University, Series: Mathematical Modelling, Programming and Computer Software*, vol 13 Issue 3 pp 17-28

[14] Dzyubenko B V, Kuzma-Kichta Yu A, Leont'ev A I, Fedik I I and Kholpanov L P 2008 Intensification of heat and mass transfer at macro-, micro- and nanoscale / B.V. Dzyubenko, - Monograph. - Moscow: FSUE "TSNIIATOMINFORM" 532 p

[15] Rohachev V A, Terekh O M, Baranyuk A V, Nikolaenko Yu E, Zhukova Yu V and Rudenko A I 2020 Heataerodynamic efficiency of small size heat transfer surfaces for cooling thermally loaded electronic components *Thermal Science and Engineering Progress* vol 20 100726

[16] Xin Jin, Beibei Shen, Hongbin Yan, Bengt Sunden and Gongnan Xie 2018 Comparative evaluations of thermofluidic characteristics of sandwichpanels with X-lattice and Pyramidal-lattice cores *International Journal of Heat and Mass Transfer* vol 127 Part B pp 268-28

[17] Konovalov D A, Ryazhskikh V I, Lazarenko I N and Kozhukhov N N 2019 Model of Cooling of Compact Surfaces by Microchannel Recuperative Heat Exchangers with a Matrix of Filamentary Silicon Single Crystals *Journal of Engineering Physics and Thermophysics*, vol 92 pp 355–64

[18] Soloveva O V, Soloveva S A and Popkova O S 2018 Modeling of the Three-Dimensional Structure of Open Cell Foam and Analysis of the Model Quality Using the Example of Pressure Drop Calculation. SCIENTIFIC NOTES OF KAZAN UNIVERSITY. SERIES OF PHYSICAL AND MATHEMATICAL SCIENCES (Proceedings of Kazan University. Physics and Mathematics Series), vol 160 No4 pp 681–94

[19] Yudai Nishiyama, Yusuke Kuwata and Kazuhiko Suga 2020 Direct numerical simulation of turbulent heat transfer over fully resolved anisotropic porous structures *International Journal of Heat and Fluid Flow* vol 81 108515

[20] Yuanzhou Zheng, Sheikholeslami M, Bhatti M M, Ahmad Shafee and Houman Babazadeh 2020 Heat transfer of nano powder inside a porous tank with use of non-equilibrium approach *International Communications in Heat and Mass Transfer* vol 116 104710

[21] Ehsan Gholamalizadeh, Ebrahim Hosseini, Mohammadreza Babaei Jamnani, Ali Amiri, Ali Dehghan saee and Ashkan Alimoradi 2019 Study of intensification of the heat transfer in helically coiled tube heat exchangers via coiled wire inserts *International Journal of Thermal Sciences* vol 14 pp 72-83

Simulation of combustion of methane-air mixture in two-dimensional approximation

K M Moiseeva and A Yu Krainov

Tomsk State University, Tomsk, 634050 Russia

Moiseeva_KM@t-sk.ru

Abstract. The paper presents a mathematical model and the results of a numerical study of the flame propagation of a methane-air mixture. The physical and mathematical formulation of the problem takes into account the thermal expansion of the gas and its subsequent movement. The problem was solved numerically using the Van Leer method to determine the fluxes at the boundaries of the computational cells. A study of flame propagation in a methane-air mixture with a methane content less than or equal to stoichiometric has been carried out. The conditions for focal ignition of a reactive mixture are determined. The influence of the channel walls on the features of flame propagation is shown. The necessity of taking into account the non-isobaricity of the combustion process at the initial stage is demonstrated.

1. Introduction

The issues of ignition and combustion of reactive mixtures have been very topical for decades. Of particular interest is the numerical simulation of the initial period of development of the ignition of a reactive mixture. Depending on the external conditions and the composition of the mixture, when the ignition sources act on the reactive mixture, the modes of both ignition and development of stable combustion, and the damping of the gas mixture after a flash are possible.

There are a number of experimental works, such as [1 - 3], in which experiments were carried out on the propagation of a methane-air mixture flame in a channel. Numerical modeling of essentially unsteady flame propagation processes, especially at the initial stage, is complicated. In particular, there are questions of setting the kinetics of combustion or taking into account the physical processes that take place during the combustion of gas mixtures in a closed or semi-closed volume.

In previous works, the problems of spark ignition and combustion of a methane-air mixture, as well as a coal-methane-air mixture, were solved in a one-dimensional approximation [4, 5]. For real conditions of the onset of ignition, for example, for the conditions of coal drifts, the one-dimensional approximation may describe the picture of the initiation and development of combustion of a reactive mixture incompletely. This paper presents the results of numerical simulation of the ignition of a methane-air mixture. The study was carried out using a two-dimensional axisymmetric formulation of the problem.

The novelty of this study lies in the numerical determination of flow parameters such as pressure, velocity vector components, inhomogeneity of the temperature field at the initial stage of combustion and in the process of establishing the flame propagation mode. Using the methods of mathematical modeling, the problem in the gas-dynamic formulation was solved and the thermal expansion of the gas with its subsequent movement was taken into account. At the same time, the problem statement

Content from this work may be used under the terms of the Creative Commons Attribution 3.0 licence. Any further distribution of this work must maintain attribution to the author(s) and the title of the work, journal citation and DOI.
Published under licence by IOP Publishing Ltd

takes into account the processes of diffusion and heat conduction. This kind of study makes it possible not only to determine the stationary modes of flame propagation, but also to reproduce numerically the non-stationary process of combustion development.

2. Mathematical model

The main assumptions used in the work are based on [4, 5]. The problem of ignition and combustion of a methane-air mixture, written in cylindrical coordinates in a two-dimensional setting, is solved. The ignition source is located in the area $-x_0 \le x \le x_0$, $-r_0 \le r \le r_0$. In the area of the hearth, an increased gas temperature is set at atmospheric pressure. An exothermic chemical reaction of the second order proceeds in the gas, the first in terms of the oxidizer and the fuel. Thermal expansion and subsequent gas movement are taken into account. It was assumed that the processes of radiant heat transfer do not affect the combustion of the mixture. For the diffusion coefficients and temperature, the law of change was set, according to the temperature ratio to the power of two-thirds. To correctly determine the heat capacity of the gas for each of the mixture compositions, the gas constant was recalculated.

For the given assumptions, the problem statement has the form:
gas mass-conservation equation:

$$\frac{\partial r\rho_g}{\partial t}+\frac{\partial r\rho_g u_g}{\partial x}+\frac{\partial r\rho_g v_g}{\partial r}=0, \tag{1}$$

gas impulse-conservation equation:

$$\frac{\partial r\rho_g u_g}{\partial t}+\frac{\partial r\left(p_g+\rho_g u_g^2\right)}{\partial x}+\frac{\partial r\rho_g u_g v_g}{\partial r}=0,\quad \frac{\partial r\rho_g v_g}{\partial t}+\frac{\partial r\rho_g u_g v_g}{\partial x}+\frac{\partial r\left(p_g+\rho_g v_g^2\right)}{\partial r}=p_g, \tag{2}$$

oxygen mass-conservation equation:

$$\frac{\partial r\rho_{O2}}{\partial t}+\frac{\partial r\rho_{O2}u_g}{\partial x}+\frac{\partial r\rho_{O2}v_g}{\partial r}=\frac{\partial}{\partial x}\left(rD_g\rho_g\frac{\partial a_{ox}}{\partial x}\right)+\frac{\partial}{\partial r}\left(rD_g\rho_g\frac{\partial a_{ox}}{\partial r}\right)-r\alpha_{ch}k_0\rho_{O2}\rho_f\exp\left(-E_a/\left(R_uT_g\right)\right) \tag{3}$$

fuel mass-conservation equation:

$$\frac{\partial r\rho_f}{\partial t}+\frac{\partial r\rho_f u_g}{\partial x}+\frac{\partial r\rho_f v_g}{\partial r}=\frac{\partial}{\partial x}\left(rD_g\rho_g\frac{\partial a_f}{\partial x}\right)+\frac{\partial}{\partial r}\left(rD_g\rho_g\frac{\partial a_f}{\partial r}\right)-rk_0\rho_{O2}\rho_f\exp\left(-E_a/\left(R_uT_g\right)\right) \tag{4}$$

gas energy-conservation equation:

$$\frac{\partial r\varepsilon_g}{\partial t}+\frac{\partial ru_g\left(\varepsilon_g+p_g\right)}{\partial x}+\frac{\partial rv_g\left(\varepsilon_g+p_g\right)}{\partial r}=\frac{\partial}{\partial x}\left(r\lambda_g\frac{\partial T_g}{\partial x}\right)+\frac{\partial}{\partial r}\left(r\lambda_g\frac{\partial T_g}{\partial r}\right)+ \\ +rQ_{ch}k_0\rho_{O2}\rho_f\exp\left(-E_a/\left(R_uT_g\right)\right), \tag{5}$$

the gas equation:

$$p_g=\rho_g T_g R_g \tag{6}$$

the initial conditions:

$$u_g(x,r,0)=v_g(x,r,0)=0,\ T_g(x,r,0)=\begin{cases}T_{hor},\ -x_0\le x\le x_0,\ -r_0\le r\le r_0\\ T_b,\ x>x_0,\ r>r_0,\end{cases}, \\ \rho_g(x,r,0)=\rho_{gb},\ \rho_{O2}(x,r,0)=a_{O2b}\rho_{gb},\ \rho_f(x,r,0)=a_{fb}\rho_{gb}. \tag{7}$$

Boundary conditions on the symmetry axis, $r = 0$:

$$\frac{\partial u_g(x,0,t)}{\partial r} = \frac{\partial v_g(x,0,t)}{\partial r} = \frac{\partial \rho_{O2}(x,0,t)}{\partial r} = \frac{\partial T_g(x,0,t)}{\partial r} = \frac{\partial \rho_g(x,0,t)}{\partial r} = \frac{\partial \rho_f(x,0,t)}{\partial r} = 0. \quad (8)$$

The following initial conditions (9) are set at the inlet of the channel ($x = 0$): the enthalpy, the gas flow rate, the axial and radial velocity of gas, the mass concentration of the oxidizer and fuel in gas.

$$\frac{\partial \rho_g(0,r,t)}{\partial x} = 0,\ a_{O2}(0,r,t) = a_{O2b},\ a_f(0,r,t) = a_{fb},\ G_g(0,r,t) = 0,\ H_g = c_v T_g,\ v_g(0,r,t) = 0. \quad (9)$$

At the right boundary of the calculation area, the exit to the environment with atmospheric conditions was specified. The upper boundary corresponded to the adiabatic wall.

The main notations in (1) - (9) are usual and correspond to the studies of [1]. In equations (5) ε is the internal energy, which is determined for gas by the equation $\varepsilon_g = p_g / (\rho_g(\gamma - 1))$, where $\gamma = c_{pg}/c_{vg}$ is the adiabatic exponent. The coefficient $\alpha_{ch} = \mu_{ox}\nu_{ox}/(\mu_f \nu_f)$ is the oxygen consumption in reactions with combustible components in the gas. For other parameters and variables, the generally accepted designations are taken: R_u is the universal gas constant, u is the velocity along the axis, v is the velocity along the radius, t is the time, T is the temperature, x is the coordinate along the axis, r is the coordinate along the radius, p is the pressure, k_0 is the rate constant of a chemical reaction, E_a is the activation energy, ρ is the density, ρ_{CH4} and ρ_{O2} are partial densities of methane and oxygen in the gas, Q_{ch} is the heat effect of the reaction. The notations μ_{ox}, μ_f are the molar masses of the oxidize and the combustible gas component, ν_{ox}, ν_f are the stoichiometric reaction coefficients.

3. Results and discussion

Simultaneous consideration of the thermal conductivity and diffusion of the gas, together with the non-isobaricity of the ignition and combustion process, leads to severe restrictions on the computational algorithm and the steps of the difference grid. The method for solving the problem, in contrast to [4], is based on the use of the Van Leer method for calculating the flows through the boundaries of the cells [6]. The Van Leer method, in contrast to Godunov method [7], more quickly determines the relationships on the faces of the computational cells. Earlier, we solved the problem of combustion of a gas suspension without taking into account the processes of heat conduction and diffusion. This made it possible to select the steps of the difference grid in space of the order of 10^{-3} *m*. In reality, the transfer processes due to the mechanisms of diffusion and heat conduction can play a significant role. The mechanisms of thermal conductivity and diffusion are especially important at the initial stage of ignition, since they give rise to gas heating and diffusion of components in the vicinity of the ignition source. Taking into account the processes of diffusion and thermal conductivity gives an estimate of the circuit diffusion of the order of 10^{-5} *m*. In this work, we took this into account. For the calculation, a difference grid with a step $h_x = h_r = 3 \cdot 10^{-5}\ m$ was taken. The time step was determined from the Courant-Friedrichs-Levy stability condition.

The characteristics of the methane-air mixture were set similarly to [1]. The ignition source was set in the range $x_0 = 6 \cdot 10^{-3}\, m$, $r_0 = 2 \cdot 10^{-3}\, m$. The selected size of the ignition source is sufficient to ignite mixtures from the considered range of compositions. For proof, we will assess the amount of energy stored in this source, $Q = c_v \Delta T \rho V$. Since a change in temperature leads to a corresponding change in the gas density, the product of the initial temperature and the initial gas density can be used. Substitution of values gives the value $Q \approx 27\, mJ$.

Examples of the results obtained are presented in Figures 1 – 4. For the calculation of Figure 1, a methane-air mixture with a volumetric methane content of 9% was used. In the figures, the coordinate axes are measured in meters.

Journal of Physics: Conference Series **2150** (2022) 012013 doi:10.1088/1742-6596/2150/1/012013

According to Figure 1, the combustion front, when gas is initiated in the center of the channel, behaves as follows. The flame expands towards the channel walls and simultaneously moves along the axis. As it progresses, the flame is drawn along the channel. Since the walls of the channel are adiabatic, the heating of the gas in the region of the walls is slower than propagation along the channel towards the outlet. After the flame reaches the exit from the channel, the combustion rate decreases and tends to the values known from the scientific literature.

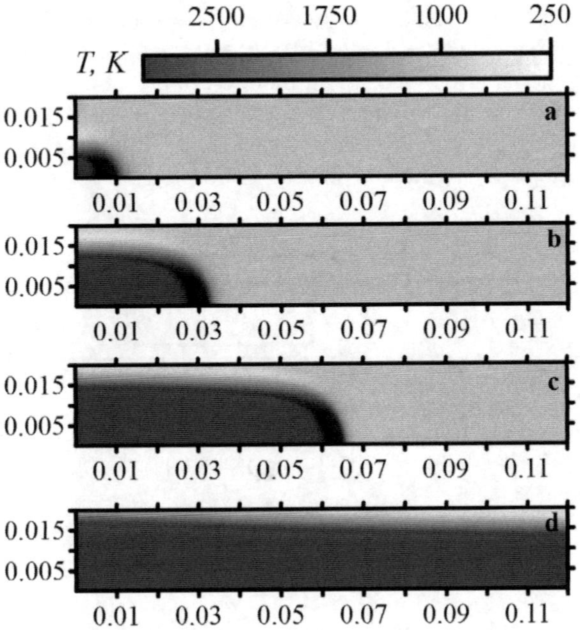

Figure 1. Gas temperature field at times $t = 10^{-3}$ s (a), $2\cdot10^{-3}$ s (b), $2.5\cdot10^{-3}$ s (c), $3\cdot10^{-3}$ s (d).

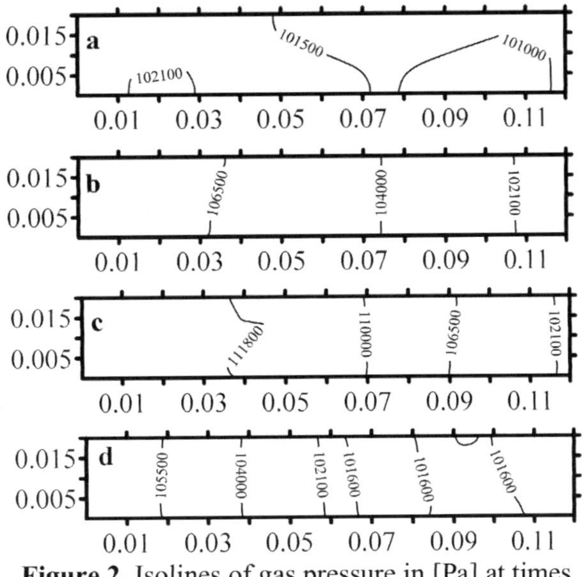

Figure 2. Isolines of gas pressure in [Pa] at times $t = 10^{-3}$ s (a), $2\cdot10^{-3}$ s (b), $2.5\cdot10^{-3}$ s (c), $3\cdot10^{-3}$ s (d).

The gas pressure at this time behaves as follows (Figure 2). At the initial stage, the increase in pressure from atmospheric is about 0.01 atmosphere (10000 Pa). As the flame spreads along the channel, the pressure in the front increases. In particular, for Figure 2c, the pressure increase from atmospheric was 0.11 atmospheres. After the gas reaches the outlet from the channel, the pressure in the channel is equalized, since the bulk of the gas has expanded into the environment. The demonstrated effect corresponds to an initial gas burst. Effects of this kind are visible to an outside observer at the initial stage of the development of a fire in a coal face. Despite the fact that under normal conditions, in a stationary mode, the apparent combustion rate of a methane-air mixture is of the order of several meters per second, at the stage of ignition and combustion development, the flame propagation rate can reach tens of meters per second. This is due to a change in pressure in the region of the combustion front, and is also explained by the action of the channel walls. The gas is unable to expand in the radial direction, which leads to an increase in pressure in the front region.

Figure 3 shows the distribution of the velocity vector along the channel at the moment of time $2.5 \cdot 10^{-3}$ *s*. This distribution corresponds to the figures 1c, 2c. It can be seen that there is a high velocity of gas movement, which is associated with the impossibility of the gas to expand in the radial direction.

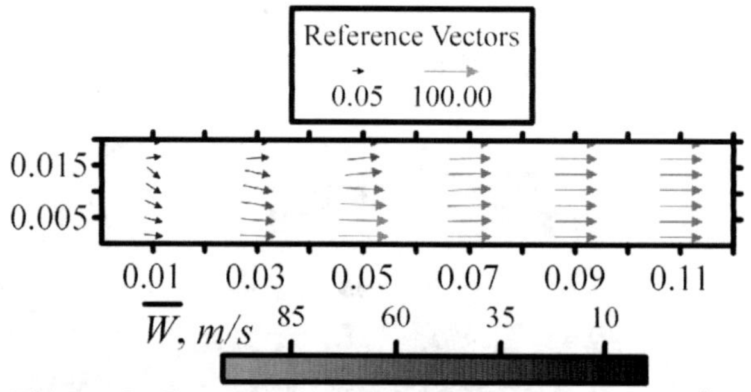

Figure 3. Gas velocity vector at time $t = 2.5 \cdot 10^{-3}$ *s*.

Figure 4 shows the dependence of the apparent velocity of flame propagation in a methane-air mixture, determined for the initial period of flame propagation. The velocity was determined from the time of flame propagation from the coordinate 0.08 *m* to the coordinate 0.09 *m*.

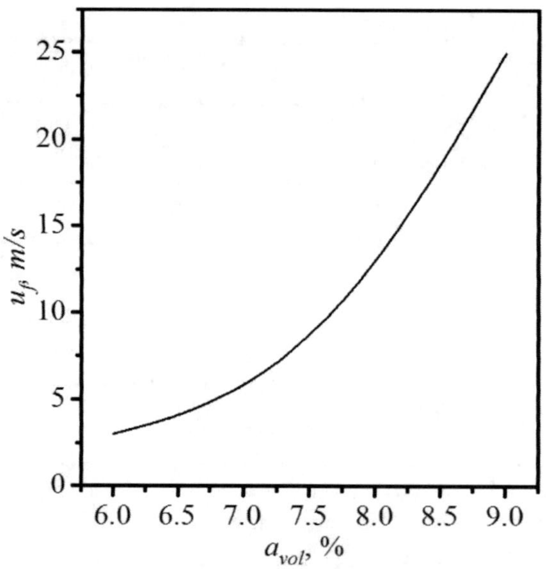

Figure 4. Dependence of the apparent combustion rate of the methane-air mixture at the initial stage of ignition on the initial methane content in the gas

The apparent burning rate of the mixture exceeds the values determined in [1]. This is because the burning rate has not yet been established. In order for the apparent combustion rate of the methane-air mixture to reach a stationary value, it is necessary to take a more extended region. The change in the normal burning rate corresponds to [1]. The normal combustion rate, defined as the difference between the apparent combustion rate and the gas velocity, is a characteristic of the composition of the mixture.

In the calculations, we also did an experiment by reducing the size of the ignition source. It was found that with a source radius less than 2 *mm* for the same volumetric content of gas ignition energy is not observed. This is due to the existence of critical values for the linear dimensions of the igniters. For example, for spark ignition problems, it is known that ignition is affected by the interelectrode spacing. In addition, it was confirmed that a mixture with a volumetric methane content of 9% ignites when the source size is an order of magnitude smaller than that required to ignite a mixture with a volumetric methane content of 7%.

4. Conclusions

A physical and mathematical model of the combustion of a methane-air mixture in a two-dimensional axisymmetric approximation has been developed. It is shown that when modeling the problem of flame propagation in a channel, it is necessary to take into account the non-isobaric nature of the process. The effects of an increase in pressure in the region of the combustion front, associated with the impossibility of gas expansion in the radial direction, are demonstrated. From the calculation, the apparent burning rate at the stage of development was determined. The resulting combustion rate exceeds the values determined for steady state flame propagation. This is due to the fact that the stage of ignition and development of combustion is accompanied by a significant change in pressure in the area of the focus. And even for an open volume, we have fairly high pressure drops. The results obtained will be used to develop a model of flame propagation through channels. The demonstrated regularities confirm that for non-stationary processes it is important to take into account the pressure change during ignition of the mixture under conditions of limited geometry along one of the sections.

Acknowledgments
This work was supported by the Russian Science Foundation (grant 21-71-10034).

References:

[1] Shchetinkov E S 1965 *The physics of combustion of gases* (Nauka, Moscow)
[2] Troshin K Ya, Borisov A A, Rakhmetov A N, etc 2013 *Russ. J. Phys. Chem. B* **7(3)** 290
[3] Hermanns R T E 2007 *Laminar Burning Velocities of Methane-Hydrogen-Air Mixtures.* (Proefschrift. Technische Universiteit Eindhoven)
[4] Moiseeva K M, Krainov A Yu 2018 *Tomsk State University Journal of Mathematics and Mechanics* **56** 79
[5] Moiseeva K M, Krainov A Yu 2018 *Combust Explos Shock Waves* **54(2)** 179
[6] Van Leer B. 1982 *Eighth International Conference on Numerical Methods in Fluid Dynamics.* **170** 507
[7] Godunov S K, Zabrodin A V, etc 1976 *Numerical Solution of Multidimensional Problems of Gas Dynamics* (Nauka, Moscow)

Multicomponent mixing on the "diffusion – convection" transition boundary at elevated pressures

V N Kossov[1], O V Fedorenko[2], M Zhaneli[1] and K Mukhatova[2]

[1] Physics Department, Abai Kazakh National Pedagogical University, 13 Dostyk Avenue, Almaty 050010, Kazakhstan

[2] Institute of Experimental and Theoretical Physics, al-Farabi Kazakh National University, 71 al-Farabi Avenue, Almaty 050040, Kazakhstan

kosov_vlad_nik@list.ru, fedor23.04@mail.ru

Abstract. An experimental and theoretical study of three-component mixing at the "diffusion – convection" boundary at elevated pressures is carried out. It is shown that the pressure dependence of the dimensionless parameter α, defined as the ratio of the experimental values of the component concentrations to those calculated by the Stefan-Maxwell equations, has characteristic regions due to the interaction of structural formations moving towards each other, in which a transition from one critical motion to another occurs. Within the framework of a linear analysis of the stability of a ternary gas mixture for a vertical circular cylinder channel, it is shown that scale perturbations determining the transition from one type of flow to another correspond to a certain value of the perturbation mode n and the critical Rayleigh numbers.

1. Introduction

Significant increase in the rate of partial mixing of the components is observed when the system passes from the mode of molecular diffusion to the area of concentration gravitational convection, under certain conditions [1, 2]. This feature is associated with the occurrence of instability of the mechanical equilibrium of the mixture and subsequent development of convective flows. Various types of convective motions in one-component medium with a monotonic dependence of density on temperature are described in detail in [3, 4]. In these works, approaches related to studies of natural gravitational convection in vertical channels are generalized for non-isothermal conditions. In addition, recommendations are presented for determining the main parameters of heat and mass transfer, related to the properties of the substance under study, which allows predicting the mixing modes under specific conditions. Moreover, it is noted that, depending on the direction of the density gradient of the mixture, various mechanisms of mixing of the components, both diffusional and caused by convection, can occur. However, propagation of the above approaches to describe the "diffusion – convection" transition in multicomponent systems in the presence of several concentration and temperature gradients requires a certain correction, since the features that are absent in binary systems are not taken into account. First of all, these include the fact of destabilizing effect of diffusion on the development of convective flows in mixtures [5-7]. The resulting movement is an important mechanism for intensifying heat and mass transfer and selective separation of components. Experimental research carried out in [8, 9] shows that even in the case of isothermal multicomponent mixing, there may be convective separation effects associated with the difference in the diffusion coefficients of the components, pressure and composition. An important feature of this type of mixing is its realizability

Content from this work may be used under the terms of the Creative Commons Attribution 3.0 licence. Any further distribution of this work must maintain attribution to the author(s) and the title of the work, journal citation and DOI.
Published under licence by IOP Publishing Ltd

under conditions of a decrease in density of the mixture with height. Therefore, solutions related to separation mechanisms in multicomponent gas mixtures are relevant, since they allow obtaining new information on the problems of combined mixing.

The aim of the work is to experimentally and numerically determine the boundary of changes in the "diffusion – concentration gravitational convection" modes in an isothermal three-component gas mixture. In terms of the diffusion Rayleigh numbers, it is supposed to obtain a family of boundary lines of changes in kinetic regimes for different modes of perturbations. The obtained results of the numerical study are compared with the experimental data.

2. Experimental data

Partial transfer of components is studied on an experimental device implementing the method of two flasks connected by a vertical cylindrical channel under isothermal conditions. The cylindrical channel has the following characteristics: $V_1 = V_2 = (76.2 \pm 0.5)\cdot 10^{-6}$ m^3, $L = (70.00 \pm 0.05)\cdot 10^{-3}$ m, and $d = (4.00 \pm 0.02)\cdot 10^{-3}$ m. Experiments are carried out in the pressure range from 0.2 to 6.0 MPa at a temperature of 298.0 K. Experimental stand and working procedure were described in detail in [10], so only the main stages of the research will be explained in detail below. The upper and lower flasks of the apparatus are filled with the studied gas mixtures up to the pressure of the experiment. Then, with the special device connecting the flasks, the channel is opened and at the same time the start time of the mixing process is fixed. At the end of experiment, the channel is blocked and the end time of mixing is fixed. Analysis of gas mixtures from each flask is carried out on a chromatograph. Experimental concentrations are normalized to the values calculated under the assumption of diffusion according to the Stefan-Maxwell equations [11]. Dimensionless parameters α_i obtained in this way characterize the corresponding type of mixing. If $\alpha_i \approx 1$, diffusion occurs. Analysis of the functional dependence of α_i on p shows, that for the system of 0.5143 He + 0.4857 Ar – 0.5148 CH_4 + 0.4852 Ar, diffusion occurs in the area of pressure $p_* \approx 1.0$ MPa (figure 1).

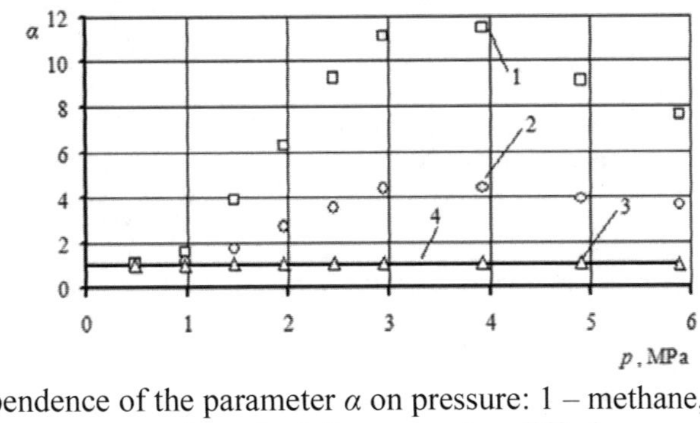

Figure 1. Dependence of the parameter α on pressure: 1 – methane, 2 – helium, 3 – argon, 4 – calculation assuming diffusion.

Further increase of pressure shows, that the parameter α_i increases. In the system under research gravitational concentration convection arises due to the instability of the mechanical equilibrium of the mixture and the pressure p_* determines the change of the "diffusion – concentration gravitational convection" regimes. At pressures significantly exceeding p_*, the linear dependence is violated.

3. Problem formulation and numerical stability analysis

Mathematical description is based on the analysis of the system of equations of continuum mechanics for multicomponent systems with respect to small perturbations [3, 4]. Macroscopic motion of an isothermal ternary gas mixture is described by a general system of hydrodynamic equations, which includes the Navier-Stokes equations, the conservation equations of the particle number of the mixture and components.

Taking into account the independent diffusion condition, under which for an isothermal gas mixture $\sum_{i=1}^{3}\mathbf{j}_i = 0;\ \sum_{i=1}^{3} c_i = 1$, this system of equations has the following form [12]:

$$\rho\left[\frac{\partial \mathbf{u}}{\partial t} + (\mathbf{u}\nabla\mathbf{u})\right] = -\nabla p + \eta\nabla^2\mathbf{u} + \left(\frac{\eta}{3} + \xi\right)\nabla\operatorname{div}\mathbf{u} + \rho\mathbf{g}$$

$$\frac{\partial n}{\partial t} + \operatorname{div}(n\mathbf{v}) = 0 \qquad \frac{\partial c_i}{\partial t} + \mathbf{v}\nabla c_i = -\operatorname{div}\mathbf{j}_i \tag{1}$$

$$\mathbf{j}_1 = -\left(D_{11}^*\nabla c_1 + D_{12}^*\nabla c_2\right) \qquad \mathbf{j}_2 = -\left(D_{21}^*\nabla c_1 + D_{22}^*\nabla c_2\right)$$

where ρ is the density; p is the pressure; n is the numerical density; t is the time; c_i is the concentration of the ith component; $\mathbf{u}$ is the mass-average velocity; $\mathbf{v}$ is the number-average velocity; $\mathbf{j}_i$ is the diffusion flux density of the ith component; $\mathbf{g}$ is the acceleration of gravity; η and ξ are the coefficients for the shearing and volume viscosities, respectively; D_{ij}^* is the practical diffusion coefficient defined using interdiffusion coefficients D_{ij}:

$$D_{11}^* = \frac{D_{13}\left[c_1 D_{32} + (c_2 + c_3)D_{12}\right]}{D} \qquad D_{12}^* = -\frac{c_1 D_{23}(D_{12} - D_{13})}{D}$$

$$D_{22}^* = \frac{D_{23}\left[c_2 D_{13} + (c_1 + c_3)D_{12}\right]}{D} \qquad D_{21}^* = -\frac{c_2 D_{13}(D_{12} - D_{23})}{D}$$

$$D = c_1 D_{23} + c_2 D_{13} + c_3 D_{12}$$

Equations (1) are supplemented by the state equation of medium

$$\rho = \rho(c_1, c_2, p) \qquad T = \text{const}$$

When solving the system of equations (1), the method of small perturbations is used [3, 4], assuming the concentration of the ith component c_i and the pressure p to be represented as follows:

$$c_i = \langle c_i\rangle + c_i' \qquad p = \langle p\rangle + p'$$

where $< c_i >$, $<p>$ are the constant average values taken as the starting point.

Considering that for $L \gg r$ (L and r are the length and radius of the diffusion channel, respectively), the differences between perturbations of the number-average $\mathbf{v}$ and the mass-average $\mathbf{u}$ velocities in the Navier-Stokes equation will be insignificant [12]. Assuming that the nonstationary perturbations of mechanical equilibrium are small, neglecting the quadratic terms on perturbations, and choosing the appropriate scales of the units of measurement (d for distance, d^2/v for time, $\rho_0 v D_{22}^*/d^2$ for pressure, D_{22}^*/d for velocity and $A_i d$ for concentration of the ith component), one can obtain a system of equations of gravitational concentration convection for the perturbed values in dimensionless quantities (strokes are omitted):

$$\mathrm{Pr}_{22}\frac{\partial c_1}{\partial t} - (\mathbf{u}\mathbf{e}_z) = \tau_{11}\nabla^2 c_1 + \frac{A_2}{A_1}\tau_{12}\nabla^2 c_2$$

$$\mathrm{Pr}_{22}\frac{\partial c_2}{\partial t} - (\mathbf{u}\mathbf{e}_z) = \frac{A_1}{A_2}\tau_{21}\nabla^2 c_1 + \nabla^2 c_2 \tag{2}$$

$$\frac{\partial \mathbf{u}}{\partial t} = -\nabla p + \nabla^2\mathbf{u} + (\mathrm{Ra}_1\tau_{11}c_1 + \mathrm{Ra}_2 c_2)\mathbf{e}_z$$

$$\mathrm{div}\,\mathbf{u} = 0$$

where $\mathrm{Ra}_i = g\beta_i A_i d^4 \left(\nu D_{ii}^*\right)^{-1}$ is the Rayleigh partial number; $\beta_i = -\frac{1}{\rho_0}\left(\frac{\partial \rho}{\partial c_i}\right)_{p,T}$; $A_i\mathbf{e}_z = -\nabla c_{i0}$ (the index 0 refers to the average values); $\mathrm{Pr}_{ii} = \nu\left(D_{ii}^*\right)^{-1}$ is the Prandtl diffusion number; $\tau_{ij} = D_{ij}^*\left(D_{22}^*\right)^{-1}$ is the parameter that determines the ratio between the practical diffusion coefficients; and $\mathbf{e}_z$ is the unit vector in the direction of the axis z.

Solution of the system of equations (2) for the cylindrical channel with circular cross section can be conditionally divided into several stages. At the first stage, we obtain the equation for the concentrations of components. Then, the equation for the velocity is written. At the second stage, solving the equation for the velocity, one can obtain the spectrum of critical Rayleigh numbers for odd and even solutions. At the third stage, we determine the boundary of the "diffusion – convection" regime change in the form [13]:

$$\tau_{11}\left(1-\frac{A_2}{A_1}\tau_{12}\right)\mathrm{Ra}_1 + \left(\tau_{11}-\frac{A_1}{A_2}\tau_{21}\right)\mathrm{Ra}_2 = \gamma^4\left(\tau_{11}-\tau_{12}\tau_{21}\right) \tag{3}$$

where $\gamma = \mathrm{Ra}^{1/4}$, i.e., $\gamma = (\mathrm{Ra}_1\tau_{11}K_1 + \mathrm{Ra}_2K_2)^{1/4}$; $K_1 = \frac{\left(1-\frac{A_2}{A_1}\tau_{12}\right)}{\left(\tau_{11}-\tau_{12}\tau_{21}\right)}$, $K_2 = \frac{\left(\tau_{11}-\frac{A_1}{A_2}\tau_{21}\right)}{\left(\tau_{11}-\tau_{12}\tau_{21}\right)}$; and A_i is the partial gradient of the concentration of the ith component.

4. Numerical results

The parameter γ entered into equation (3) is determined by two indices n and l [13]. Index n defines the azimuthal structure of the movement, and index l determines the radial structure. The transition from the diffusion region to the convective one is characterized by the values of indices n and l equal to unity, which corresponds to diametrically antisymmetric motion, in which the cylinder is divided into two parts, in one of which the gas rises and in the other falls. If $n = 2$ and $l = 1$, then the cylinder is divided into four parts, along two of which the gas rises, and along the other two the gas falls, etc. Thus, various schemes of critical motions are realized in the gas mixture. The transition from one critical movement to another can be recorded experimentally, which is confirmed by the data shown in figure 1.

For the system of 0.5143 He + 0.4857 Ar – 0.5148 CH_4 + 0.4852 Ar, a stability analysis is performed based on equation (3). Figure 2 shows the partial Rayleigh numbers and critical lines for various disturbance modes.

The partial Rayleigh numbers in accordance with the system of equations (2) are calculated by the following formulas:

$$\mathrm{Ra}_1 = \frac{gnr^4\Delta m_1}{\rho\nu D_{11}^*}\cdot\frac{\partial c_1}{\partial z} \qquad \mathrm{Ra}_2 = \frac{gnr^4\Delta m_2}{\rho\nu D_{22}^*}\cdot\frac{\partial c_2}{\partial z} \tag{4}$$

where $\Delta m_1 = m_1 - m_2$, $\Delta m_2 = m_2 - m_3$, m_i is the molecular mass of ith component.

The calculated data shown in figure 2 indicate that in the system under consideration there are pressure regions at which various types of critical motions are observed. Thus, in the pressure range from 0.98 to 1.47 MPa, a transition from the diffusion mode to the convective one is observed, which corresponds to the first mode of perturbations. For $n = 1$ and $l = 1$, the critical Rayleigh numbers have the following values: $\mathrm{Ra}_1 = 65.24$, $\mathrm{Ra}_2 = 23.98$ at $\gamma_1^1 = 2.871$. In the pressure range 2.94 - 3.92 MPa, as shown in figure 1, a maximum is recorded to correspond to $n = 2$ and $l = 1$, i.e., there is a change in the type of critical motion, which corresponds to the critical Rayleigh numbers $\mathrm{Ra}_1 = 315.7$, $\mathrm{Ra}_2 =$

116.1 at $\gamma_2^1 = 4.259$. The next transition corresponding to $n = 3$ and $l = 1$ is observed in the pressure range of 4.90 - 5.89 MPa, while the cylindrical channel is divided into six parts. In the system, the scale of disturbances changes to the steady-state convective flow regime and the next convective regime arises, which corresponds to the critical Rayleigh numbers $Ra_1 = 904.8$, $Ra_2 = 332.6$ at $\gamma_3^1 = 5.541$.

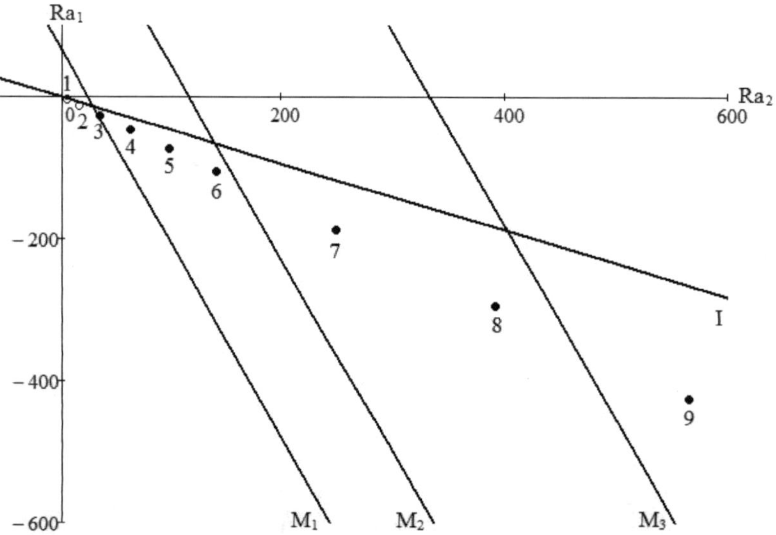

Figure 2. Stability map for the system of 0.5143 He + 0.4857 Ar – 0.5148 CH_4 + 0.4852 Ar. I is the line of zero density gradient. M_1, M_2 and M_3 are the lines of monotonic disturbances at $n = 1$, $n = 2$ and $n = 3$. Points are the partial Rayleigh numbers calculated at the following pressure values: 1 – p = 0.49; 2 – 0.98; 3 – 1.47; 4 – 1.96; 5 – 2.45; 6 – 2.94; 7 – 3.92; 8 – 4.90; 9 – 5.89 MPa.

Thus, the regions associated with nonlinear changes in the process intensity correspond to new scales of disturbances. In this case, the experimental partial Rayleigh numbers Ra_i are located on the plane of the Rayleigh numbers (Ra_1, Ra_2) near the boundary lines corresponding to different perturbation modes n.

5. Conclusions

The boundary of the "diffusion – convection" transition in a three-component gas mixture, where one of the components is uniformly dissolved in the other two under isothermal conditions and elevated pressures, has been studied experimentally and theoretically. It has been found that the parameter α, defined as the ratio of the experimental concentrations to the values of the concentrations calculated under the assumption of diffusion using the Stefan-Maxwell equations, changes nonlinearly with increasing pressure. The resulting dependence has minima and maxima, the presence of which is explained by the interaction of convective structural formations moving towards each other. For the investigated system of 0.5143 He + 0.4857 Ar – 0.5148 CH_4 + 0.4852 Ar, a linear stability analysis has been carried out for a vertical cylindrical channel with a circular cross section. For various perturbation modes, the critical Rayleigh numbers at which the type of critical convective motions changes have been determined. The results of numerical studies are shown to be in good agreement with experimental data.

Acknowledgments
The work was supported by the Science Committee of the Ministry of Education and Science of the Republic of Kazakhstan (project no. AP09259248) and the grant of Abai Kazakh National Pedagogical University (Grant No. 258-is19).

References
[1] Dil'man V V and Lotkhov V A 2015 *Theor. Found. Chem. Eng.* **49(1)** 102
[2] Zhavrin Y I, Kosov V N, Fedorenko O V and Akzholova A A 2016 *Theor. Found. Chem. Eng.* 50 171.
[3] Gershuni G Z and Zhukhovitskii E M 1976 *Convective Stability of Incompressible Fluids* (Jerusalem: Keter Publications)
[4] Nield D A and Bejan A 2006 *Convection in Porous Media* (New York: Springer)
[5] Dil'man V V, Lipatov D A, Lotkhov V A and Kaminskii V A 2005 *Theor. Found. Chem. Eng.* **39** 566
[6] Kaminskii V A 2011 *Rus. J. Phys. Chem.* A **85** 1986
[7] Ruev G A, Fedorov A V and Fomin V M 2009 *J. App. Mech. and Tech. Phys*. **50** 49
[8] Kosov V N, Seleznev V D and Zhavrin Yu I 1997 *Tech. Phys*. **42(10)** 1236
[9] Kossov V, Fedorenko O, Kalimov A and Zhussanbayeva A 2021 *Fluids* **6(5)** 177
[10] Asembaeva M K, Kosov V N, Krasikov S A and Fedorenko O V 2019 *Tech. Phys. Letters* **45(21)** 1071
[11] Bird R B, Stewart W E and Lightfoot E N 2007 *Transport Phenomena* (New York: John Wiley & Sons)
[12] Kossov V, Krasikov S and Fedorenko O 2017 *Eur. Phys. J. Spec. Top*. **226** 1177
[13] Kosov V N, Fedorenko O V, Zhavrin Y I and Mukamedenkyzy V 2014 *Tech. Phys*. **59(4)** 482

Experimental study of an internal combustion engine fueled by a low-calorific value producer gas

G I Nikitina[1,2], A N Kozlov[1] and M V Penzik[1]

[1] Melentiev Energy Systems Institute SB RAS,
130 Lermontov street, Irkutsk, 664033, Russia
[2] Ural Federal University named after the first President of Russia B.N. Yeltsin,
19 Mira street, Ekaterinburg, 620002, Russia

uge87@mail.ru

Abstract. This paper describes an experimental study of the operation of an internal combustion engine of fueled by a low-calorific value gas. The main operating parameters of low-power ICE were determined. Efficiency was also evaluated when the ICE was converted to operate on producer gas. In the experiment, it was shown that the engine reached a stable operating mode under load and data on the temperature and exhaust gases composition were obtained. According to our estimates, in the steady-state operation of the internal combustion engine with a load, the efficiency factor was about 22 %. When using the model gas, the from generator output power, was about 30-40 % of the nominal value, under variable load conditions. However, it was found that in steady-state operation, the power of the internal combustion engine was 40-55% of the nominal value.

1. Introduction

The problem of power supply for autonomous consumers [1] has been relevant for several centuries, however, with an increase in the standard of living, it is required to provide high-quality energy resources with guaranteed reliability. At the same time, it is necessary to solve the issues of environmental safety and economic efficiency in energy production even at low-power plants.

Decentralized energy supply areas often have local fuel resources, low-grade solid fuels and various wastes. One of these sources is biomass, such as wood waste from logging operations. At the same time, their conversion into high quality electrical and thermal energy can be carried out using gasification technologies, followed by the use of producer gas in internal combustion engines (ICEs). As a rule, the gas produced during air gasification has a low calorific value of the order of 4–5 MJ/m^3, which makes its transportation unprofitable [2], whereas its use in gas internal combustion engines can provide the replacement of expensive imported diesel fuel and provide the consumer with electrical energy [3].

However, there is the problem of the reduce efficiency of internal combustion engines operating on off-design fuels. Difficulties also arise with the start of engines on low-calorific value producer gas; in this case, ignition doses of diesel fuel or, for example, methane are used.

Despite the flexibility of internal combustion engines in operation with variable load, possible fluctuations in the composition of the producer gas, its contamination with dust particles, tar (up to 10 g/nm^3) [4], and high gas humidity can negatively affect the reliability and failure-free operation of ICEs.

Content from this work may be used under the terms of the Creative Commons Attribution 3.0 licence. Any further distribution of this work must maintain attribution to the author(s) and the title of the work, journal citation and DOI.
Published under licence by IOP Publishing Ltd

Therefore, after the gasification reactor, a gas purification unit for removal of solid particles and tar is required, as well as dehumidification for moisture condensation and cooling of hot gas [5].

In a previous thermal calculation of an internal combustion engine, the operating conditions were estimated for the electric power levels of 30–300 kW according to [6].

The authors and colleagues have developed a biomass conversion reactor with a power of 5 kW [7], but for a comprehensive assessment of the operation of a gasifier with an internal combustion engine, a separate study of the power part is required, therefore an experimental setup with a small-power ICE was made. The purpose of the work was to carry out an experimental study of the modes and features of ICE operation on a low-calorific value producer gas.

2. Material and methods

2.1. Experimental setup

The gasoline internal combustion engine was modified and converted to gas fuel. Test runs with a model gas were carried out using the Kronwerk 94649 LK1500 generator.

The Kronwerk generator is an autonomous power plant used to power electrical equipment in the absence of a household network or for remote consumers. The air-cooled engine with a volume of 94 cm^3 consist of a 4-stroke single-cylinder with a rated power of 1 kW and a maximum power of 1.2 kW.

Using this generator, a series of experiments were carried out with a model gas which was, a mixture of combustible gases with CO_2 and N_2. Gas was supplied from a cylinder through a system of pressure gauges; air was sucked in through a branch pipe and mixed with the gas. After that, the mixture entered the internal combustion engine, exhaust gases were emitted at the outlet, and electricity was generated and fed to the load system (Figure 1).

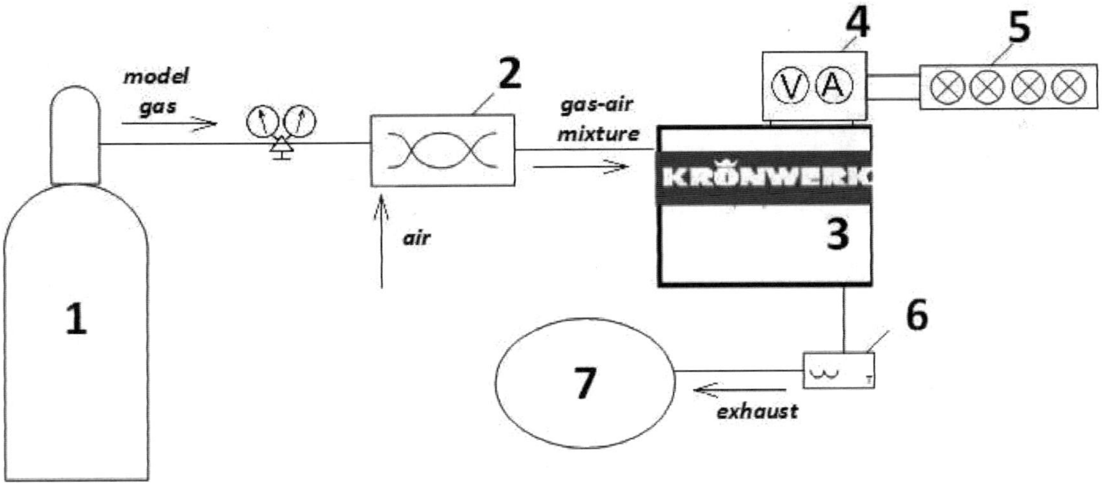

Figure 1. Schematic of the experimental setup:
1 – gas cylinder; 2 – gas mixing unit; 3 – generator; 4 – voltmeter and ammeter;
5 – load system with electric lamps; 6 – thermocouple; 7 – gas analyzer.

The temperature and composition of the exhaust gases were measured, and the parameters of electrical energy under variable load conditions were determined. The load system consisted of four 100–150 W electric lamps. To determine the composition of the exhaust gases, a TESTO 340 Gas Analyzer and an Inficon Micro GC fusion Chromatograph were used.

Journal of Physics: Conference Series **2150** (2022) 012015 doi:10.1088/1742-6596/2150/1/012015

Since the air gasification of biomass gives a producer gas with a high nitrogen content but with a low calorific value, a model gas for experimental engine runs has been produced. Its composition was as follows (molar fraction, %): C_2H_6 – 0,964; C_2H_4 – 0,959; H_2 – 17,26; CH_4 – 4,26; CO – 17,28; CO_2 – 13,09; N_2 – 46,187. This composition typical of the gas produced by air gasification of biomass.

All measured parameters were recorded over time and entered into an experimental database.

2.2. Calculation

In the calculation of the ICE efficiency, the generated power and the energy of the supplied gas were evaluated. The calorific value of the gas mixture was 6.75 MJ/m^3.

The average gas consumption in the experiments was 18 liter per minute. Thus, the amount of energy supplied to the ICE was calculated by multiplying the gas consumption by its calorific value.

The energy output from the generator (N) was determined from the readings of the voltmeter and ammeter. In steady-state operation, the power level was 325–448 W (Figure 2).

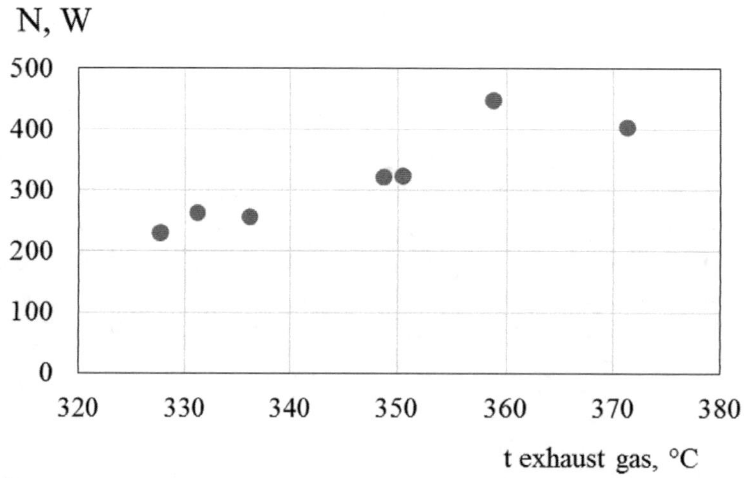

Figure 2. Dependence of power on exhaust gas temperature.

The effective efficiency factor η_e is the ratio of the amount of heat converted into effective work by the engine Le on the amount of heat Q_e supplied to perform this work.

$$\eta_e = \frac{L_e}{Q_e} \tag{1}$$

The excess air factor was calculated from the exhaust gases composition by using the oxygen formula:

$$\alpha = \frac{21}{21 - O_2} \tag{2}$$

The thermal calculation (Table 1) of the internal combustion engine was made for natural gas (methane) and the model gas. **However, for the calculation, it was assumed that the engine produces the same power level.**

Table 1. Thermal calculation results for the gas-fuelled engine.

Index	Notation	Measurement unit	Model gas	Methane
Excess air factor	α		2	2
Specific weight of combustible gas	γ_H	kg/nm^3	1.12	0.766
Lowest heating value of combustible gas	Q	kcal/nm^3	1633	8000
Volume of theoretically required air for combustion of a volume unit of fuel	L_o	nm^3/nm^3	1.22	8.95
Electric power of the engine	N_e	W_e	455	455
Specific effective fuel consumption	V_e	m^3/kW$_e$	5.66	0.637
Specific effective heat consumption	q_e	kcal/kW$_e$	7553	4327
Effective engine efficiency	η_e	%	20.5	19.6

The estimated temperature in the cylinder of the internal combustion engine during the combustion of methane was 1320 °C, of model gas was 1240 °C. The calculation was made for dry gas (disregarding external moisture). The heat from syngas cooling can be transferred through the heat exchanger to dry the fuel or can be used to heat the air supplied to the gasifier.

3. Results and discussion

According to our estimates, the internal combustion engine in steady-state operation has an efficiency factor of about 22 %. The average excess air factor is 2. Therefore, a comparison of the experimental data and thermal calculation results shows that they are in good agreement.

For example, the generator output power obtained using the model gas under variable load conditions operation was about 30–40 % of the nominal value. In steady-state operation, the power of the internal combustion engine is 40–55% of the nominal value. The calculated maximum experimental power is 455 W.

However, when comparing the thermal calculation results for the ICE on the model gas and the ICE on methane, it can be seen that the specific gas consumption is higher for the model gas than for methane.

Figure 3 shows data on the exhaust gas composition for a number of experimental points; nitrogen is about 80 percent of the total composition. The presence of residual oxygen indicates an excess of air supply.

The amount of pollutants in the exhaust gases of the internal combustion engine was determined when using the model gas. The exhaust gases were found to contain a small amount of SOx (10–15 ppm), produced by the combustion of engine oil which penetrated into the combustion chamber when the engine was started and stopped.

The NOx emissions from the combustion of the model gas in the internal combustion engine were estimated. The steady-state amount of NOx is 10–12 ppm.

Since the consumption of the model gas changed with increasing electrical load, the NOx content also changed.

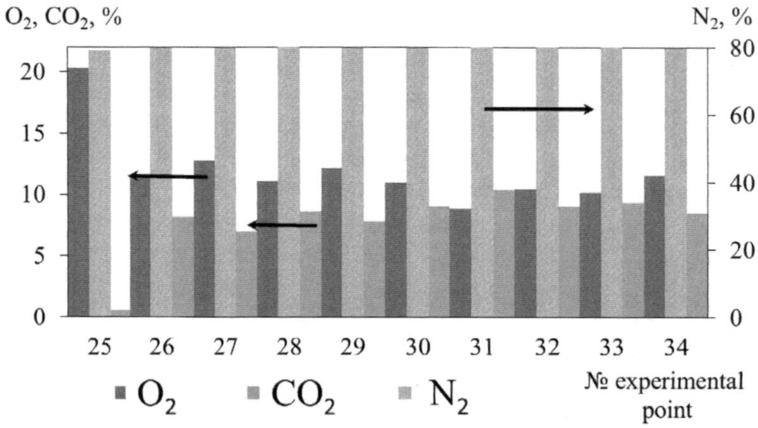

Figure 3. Exhaust gas composition.

It is shown (Figure 4) that after the engine warms up, the exhaust gas temperature reaches a constant value of ~350 °C and does not depend on the load and gas consumption.

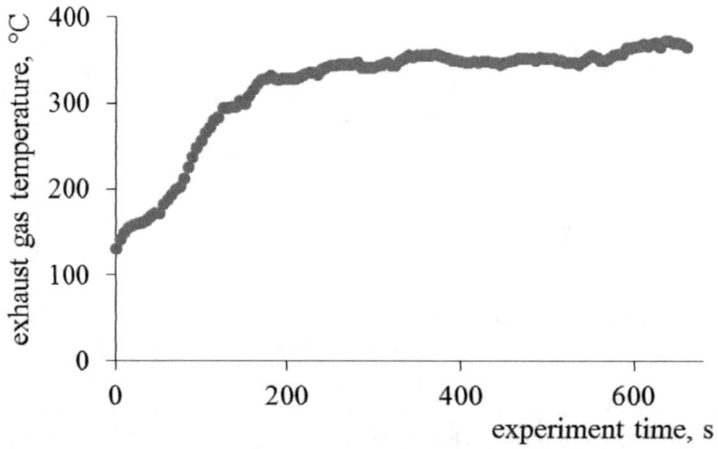

Figure 4. Dependence of the exhaust gas temperature on the time of the experiment.

4. Conclusions

According to the results of the experimental study using the model gas, satisfactory operating conditions of the internal combustion engine under load were worked out. Data on the output power, temperature and composition of exhaust gases were obtained.

The generator output power, which obtained using the model gas at a voltage of 220±20V was about 30–40 % of the nominal value, under variable load conditions. However, it was found that in the steady state the power of the internal combustion engine is 40–55% of the nominal value. According to our estimates, in the steady-state operation of the internal combustion engine with a load, the efficiency factor is about 22 %, which corresponds to the calculated values.

The data on the exhaust gas temperature in the range of 320–370 °C at the calculated power level and the presence of residual oxygen indicate an excess of supplied air. In this connection, it is planned to modify the supply channel.

The results of the study will make it possible to further improve the operation of the internal combustion engine on a ow-calorific value producer gas, which will be directly obtained by gasification of various solid fuels. The next stage of the study work is the launch of a biomass conversion reactor with this internal combustion engine.

Acknowledgements
This reported study was funded by the Russian Foundation for Basic Research (RFBR) No. 19-58-80016; the Department of Science and Technology of India (DST) No. CRG/2018/004610, DST/TDT/TDP-011/2017; the Ministry of Science and Technology of the People's Republic of China (MOST) No. 2018YFE0183600; the National Research Council of Brazil (CNPq) No. 402849/2019-1; and the National Research Foundation of South Africa (NRF) No. BRIC190321424123; the resources of the High-Temperature Circuit Multi-Access Research Center were used (Ministry of Science and Higher Education of the Russian Federation, Project No. 13.CKP.21.0038).

References
[1] Banerjee R 2006 *Energy Policy* **34** 101
[2] Kozlov A N, Tomin N V, Sidorov D N, Lora E E S, Kurbatsky V G 2020 *Energies* **13** 2632
[3] Arun P 2015 *International J. of Renewable En-ergy Research* **5** 3
[4] Bocci E, Sisinni M, Moneti M, Vecchione L, DiCarlo A, Villarini M 2014 *Energy Procedia* **45** 247
[5] Asadullah M 2014 *Renewable and Sustainable Energy Reviews* **29** 201
[6] Khudyakova G I, Kozlov A N, Svishchev D A 2017 *J. of physics: conference series* **891** 012200
[7] Svishchev D A, Kozlov A N, Penzik M V 2019 *Energy Procedia* **158** 649

Journal of Physics: Conference Series **2150** (2022) 012016 doi:10.1088/1742-6596/2150/1/012016

Development of the design of a new working section cooled by a dispersed coolant flow for a stand with induction heating

A V Dedov, A T Komov, V S Shteling, P P Shcherbakov, V V Iliyn and A V Zakharenkov

General Physics and Nuclear Fusion Department, National Research University "Moscow Power Engineering Institute", 14 Krasnokazarmennaya Street, Moscow, Russia

Schteling@gmail.com

Abstract. The paper presents a description of the design of a new working section cooled by a dispersed coolant flow for a stand with induction heating. The choice of the material of the working area and the method of spraying has been substantiated. The first commissioning experiments were carried out in a stationary cooling mode, with the parameters of the coolant p = 1.7 · 105 Pa, G = 3.8 · 10^{-3} kg/s.

1. Introduction

In modern technology, it is often necessary to remove heat fluxes with an energy density of several MW/m^2. This is typical for such areas as aerospace and nuclear technology, microelectronics, and energy. At the same time, the cooled objects are very diverse in terms of their weight and size parameters: from compact ones, which are found in computer and laser technology, to bulky and voluminous ones, such as, for example, rolled products of the steel industry during quenching and cooling. One of the most promising ways to cool a high-temperature surface is to use a dispersed coolant.

Obtaining a polydisperse mixture of droplets is ensured through the use of nozzles - devices that generate a dispersed flow of the coolant, while it is necessary to achieve a uniform distribution of the liquid over the cross section of the spray plume. Cooling with a dispersed coolant has proven itself in power engineering, mechanical engineering, metallurgy, but its application is not limited only to these areas of industry. Due to its high efficiency and relative ease of implementation, it may well be applicable for cooling the structures of modern thermonuclear power systems exposed to high-density thermal loads.

The process of heat transfer during cooling by sputtering is poorly understood: complex physics, involving the simultaneous interaction of sputtered droplets with a cooled surface, their evaporation, convection and vapor condensation [1, 2]. Models are known that describe these complex processes [3-5], but it is hardly possible to consider them exhaustive. That is why experimental research in this area is of great relevance.

The most detailed review of all relevant works on the topic of dispersed cooling is presented in [6].

The purpose of this work is to study the cooling of a high-temperature surface by a dispersed stream of water, on an experimental bench that simulates the energy-loaded elements of thermonuclear installations.

Content from this work may be used under the terms of the Creative Commons Attribution 3.0 licence. Any further distribution of this work must maintain attribution to the author(s) and the title of the work, journal citation and DOI.
Published under licence by IOP Publishing Ltd

2. Description of the experimental stand

To cool the heated target, a closed hydraulic circuit was designed and installed (Fig. 1). Distilled water is used as a heat carrier, which is circulated by a pumping system (3). To control the temperature of the heated target, 6 thermocouples are installed. The L-CARD information collection system (4) is connected to thermocouple sensors, a digital flow meter (5) and a pressure meter (6). The scheme provides for the necessary fittings for regulating flow and pressure, measuring them, as well as a condenser for the formed water vapor.

Tanks 1 and 2 with level gauges are designed to measure the flow rate of the evaporated and non-evaporated parts of the dispersed stream.

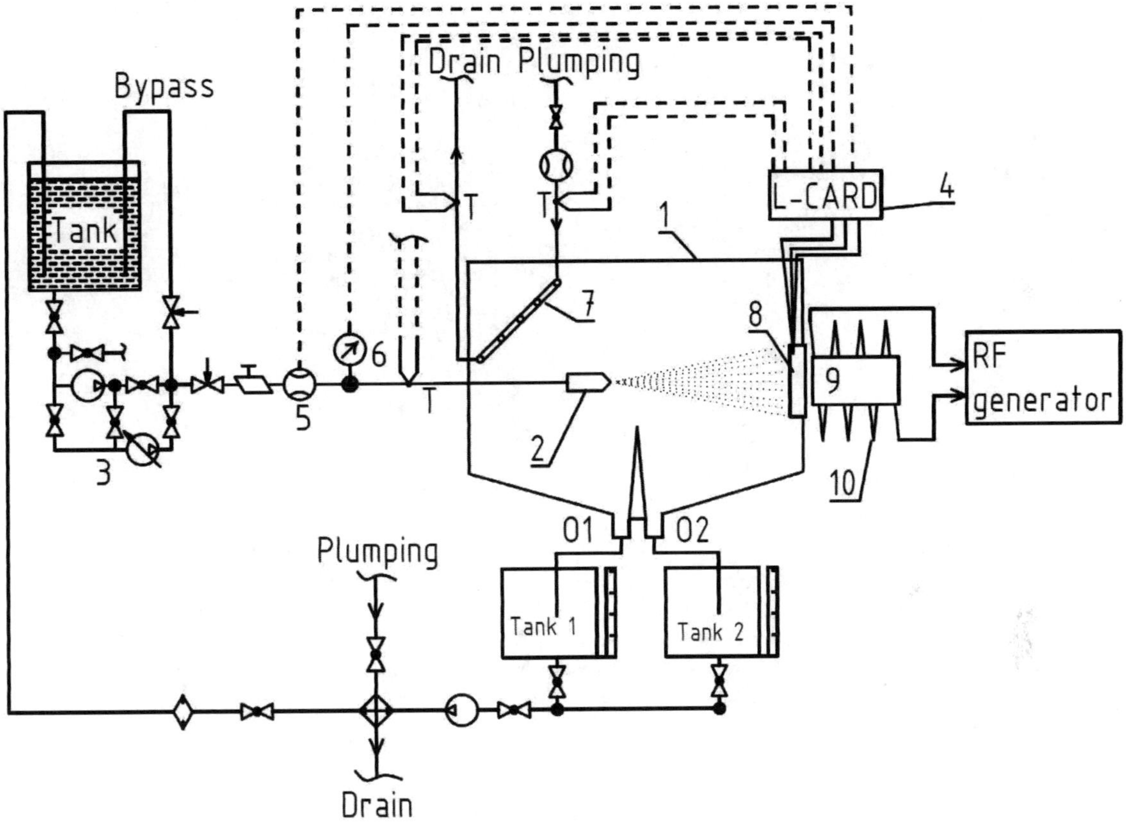

Figure 1. Schematic diagram of the stand: 1 – experimental chamber, 2 – nozzle, 3 – pumping system, 4 – L-CARD information collection system, 5 – pressure meter, 6 – digital flow meter, 7 – condenser, 8 – target, 9 – ferrite core, 10 – induction coil

3. Description of the induction heating circuit of the target

A nickel cylinder 50 mm in diameter and 10 mm thick (Fig. 2) is a heated element that is attached to the chamber wall using a magnetically transparent ceramic flange with low thermal conductivity.

Between the flange and the chamber wall there is a graphite o-ring, which has sufficient ductility and is resistant to high temperatures, which fully satisfies the experimental conditions.

On the outside of the heated element there is an induction coil connected to the RF generator and the cooling system. In order to increase the inductance, the coil is double wound. A magnetic concentrator in the form of a cylinder, made of ferrite 10000NM, is placed in the coil. This ferrite core allows for efficient one-sided heating with high energy density.

The choice of nickel as a material for the heated element is due to its ferromagnetic properties, high Curie temperature (~ 358°C), relatively high thermal conductivity (λ = 91 W/(m · K) at a temperature of 20°C), as well as corrosion resistance [7]. This working section, due to its small thickness, has a low thermal inertia, which makes it possible to obtain the highest quality results when studying a stationary cooling mode with a dispersed flow.

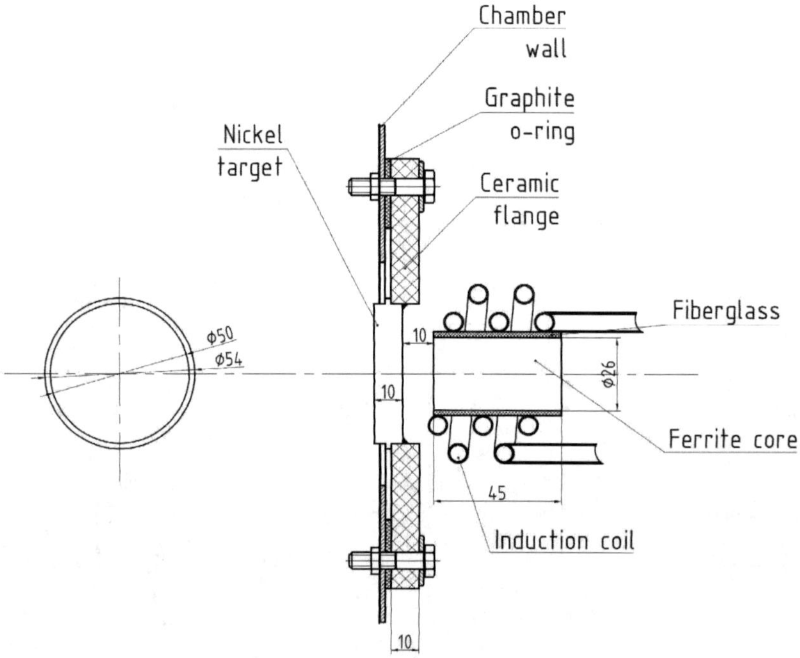

Figure 2. Induction heating circuit for a nickel target

Below is a layout of thermocouples in a nickel target (Fig. 3), for the possibility of calculating heat fluxes and controlling the rate of cooling, as well as getting an idea of its uniformity. The temperature field in the target is measured by six thermocouples in two sections. In each of the two sections, three thermocouples are installed, located in the target radially at different depths. Cable thermocouples with a diameter of 1 mm are used.

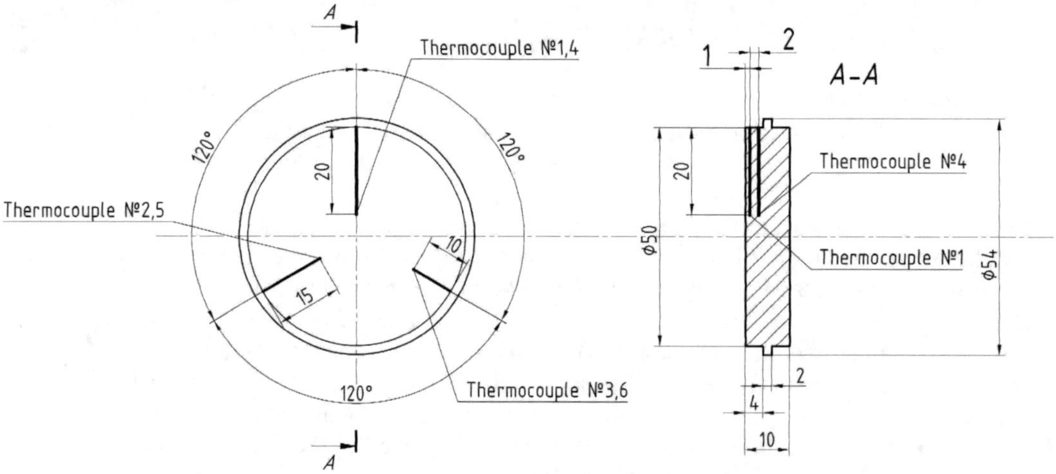

Figure 3. Layout of thermocouples in the target

4. Hydraulic nozzle

A hydraulic centrifugal nozzle from Fluidics (Fig. 4) is investigated as a dispersed flow generator as the most economical in terms of energy consumption for spraying a liquid [8] with a diameter of 0.7 mm. The average droplet diameter declared by the manufacturer is 35 microns.

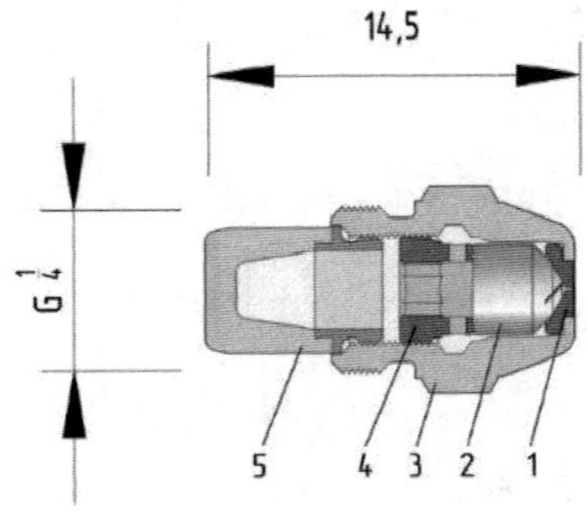

Figure 4. Nozzle with conical insert: 1 – nozzle (Ø 0.7 mm), 2 – swirler, 3 – body, 4 – limiter, 5 – filter

5. Technique of the experiment

The procedure for conducting an experiment on thermal stabilization in a stationary mode with a dispersed flow is as follows:

1) setting the required parameters for the pressure and flow rate of the coolant;
2) setting the required power of the RF generator;
3) after establishing a stationary temperature distribution in the target for 60 seconds, at a frequency of 10 Hz, the values of the pressure and flow rate of the coolant, and the target temperatures are recorded. The maximum temperature deviation in stationary mode was no more than 2°C. Subject to this condition, a stationary mode was recorded.

An example of recording measurements in an experiment with stationary modes selected on them at different powers is shown in Fig. 5, which shows the readings of two thermocouples located in different coordinates along the target thickness for different operating parameters.

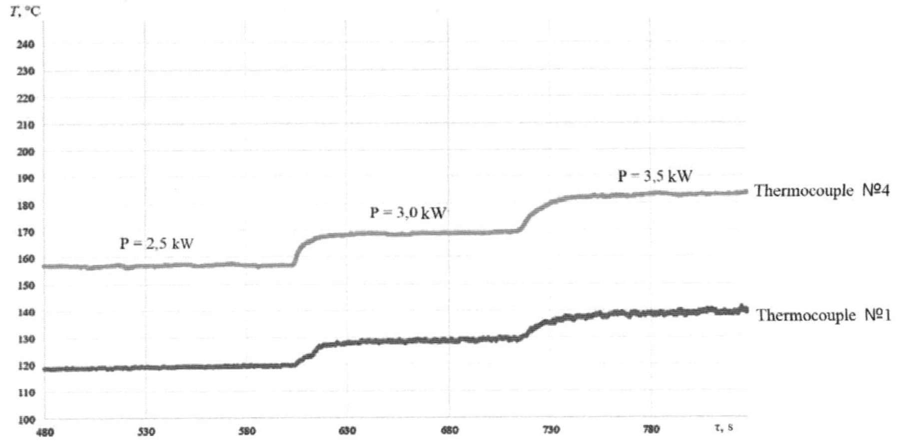

Figure 5. From the experiment protocol: coolant parameters p=1.7·10^5 Pa, G=3.8·10^{-3} kg / s , —, — – readings of thermocouples T1 and T4, respectively

6. Technique for processing primary experimental data

Primary experimental data processing scheme consists of:

1. Time averaging of the readings of thermocouples, which measure the temperature of the working section and the coolant flow in a stationary mode, is performed.
2. Target temperatures calculation averaged over the sections $x_1 = 1$ mm and $x_2 = 3$ mm (see Fig. 3), using the temperatures averaged over time in step 1 $\bar{T}_1 = \frac{T_1+T_2+T_3}{3}$; $\bar{T}_2 = \frac{T_4+T_5+T_6}{3}$.
3. Since in the considered temperature range the thermal conductivity along the thickness of the target changes insignificantly, $\lambda_t \approx$const can be considered, while the temperature field is described by a linear dependence, and the wall temperature, determined by the extrapolation method (see Fig. 6).

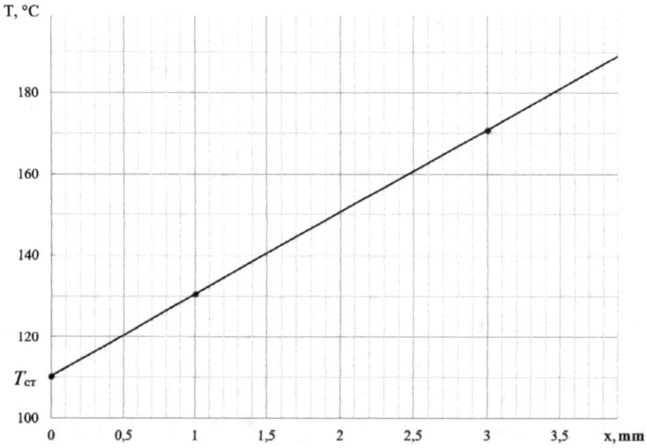

Figure 6. Temperature distribution over the target thickness

4. Heat flux density calculation for a given stationary mode $= \lambda \frac{\bar{T}_2 - \bar{T}_1}{x_2 - x_1}$, according to Fourier law.
5. Calculation of the heat transfer coefficients according to the Newton-Richman law: $\alpha = \frac{q}{\Delta T_S} = \frac{q}{T_w - T_S}$.

Table 1 shows the primary data of adjustment experiments averaged over time for each stationary mode. Table 2 shows the calculated thermophysical values.

Table 1. Primary data of adjustment experiments

$p \cdot 10^{-5}$, Pa	$G \cdot 10^{6}$, m^3/s	P, kW	T_1, °C x=1 mm R=20 mm	T_2, °C x=1 mm R=15 mm	T_3, °C x=1 mm R=10 mm	T_4, °C x=3 mm R=20 mm	T_5, °C x=3 mm R=15 mm	T_6, °C x=3 mm R=10 mm
		2.5	121.12	122.84	121.05	156.67	154.38	153.74
		3.0	131.32	130.42	129.55	170.12	172.16	170.03
		3.5	138.21	137.4	132.54	182.31	180.96	178.74
1.7	3.8	4.0	143.05	138.96	135.92	190.51	187.37	181.53
		4.5	149.27	146.91	141.43	203.21	196.08	192.97
		5.0	158.30	153.03	150.85	214.64	210.27	204.28
		5.5	166.47	161.2	158.03	226.39	220.47	215.93

Table 2. Calculated thermophysical values

$p \cdot 10^{-5}$, Pa	$G \cdot 10^{6}$, m^3/s	P, kW	$\bar{T}_1$, °C	$\bar{T}_2$, °C	$T_{ст}$, °C	q, kW / m^2	α, kW / ($m^2 \cdot K$)
1.7	3.8	2.5	121.67	154.93	111.1	1370.0	271.9
		3.0	130.43	170.77	113.1	1636.8	159.5
		3.5	136.05	180.67	116.8	1793.2	130.4
		4.0	139.31	186.47	118.2	1885.1	119.9
		4.5	145.87	197.42	118.9	2038.0	101.4
		5.0	154.06	209.73	119.9	2172.3	82.8
		5.5	161.90	220.93	120.7	2275.9	70.3

References

[1] Kim J. Spray cooling heat transfer: the state of the art. Int J Heat Fluid Flow 2007;28:753–67.
[2] Wen-Long Cheng, Wei-Wei Zhang, Hua Chen, Lei Hu, Spray cooling and flash evaporation cooling: The current development and application, Renewable and Sustainable Energy Reviews, Volume 55, 2016, Pages 614-628
[3] Youssef, M.R. Modeling the effect of a spray on a liquid film on a heated surface, PhD Thesis, Dept. of Civile and Environmental Engineer, West Virginia University, 2007.
[4] Zhao, R., Cheng, W., Liu, Q., Fan, H. Study on heat transfer performance of spray cooling, model and analysis, Heat Mass Transfer, Vol. 46, pp. 821–829, 2010.
[5] I. V. Kazachkov. Heat transfer and dynamics of the droplet on a superheated surface, Heat Mass Transfer, Issue 2, Volume 7, pp. 47-57, 2012.
[6] G. Liang, I. Mudawar. Review of spray cooling – Part 1: Single-phase and nucleate boiling regimes, and critical heat flux // International Journal of Heat and Mass Transfer, 115 (2017), pp. 1174–1205
[7] Mechanical engineering. Encyclopedia. Volume II-3. Non-ferrous metals and alloys. Composite metallic materials / Collective of the Russian Academy of Sciences, Moscow: Mashinostroenie, 2013. - 784 p.
[8] Dityakin Yu.F. Spraying liquids / Yu.F. Dityakin, L.A. Klyachko - M .: Mashinostroenie, 1977.- 206 p.

Design Optimization of a Helical Coil Gas Cooler Based on the Results of CFD Modeling of Erosion Wear

V A Mikula, G E Maslennikov and T F Bogatova

Ural Federal University, Yekaterinburg, Russia

tb-upi@mail.ru

Abstract. Simulation of erosion wear and design optimization have been performed for a convective gas cooler with a helical coil. Based on the results of simulation of the standard gas cooler design with a flat baffle used in Shell gasification-based combined cycle unit, it is concluded that the particle impact angle is the main factor determining the erosion maximum. To reduce erosion, it is necessary to install a structural element instead of the flat baffle to align the flow path of ash particles at the inlet to the gas cooler. The results of simulation for various baffle shapes show that a hemispherical baffle is optimal. The use of a hemispherical baffle plate made it possible to align the ash particle flow path at the inlet to the gas cooler channels and reduce the maximum level of erosion by a factor of almost 4 compared to the standard geometry of the baffle plate.

The erosion wear of the surfaces of heat exchangers in two-phase flow is a fairly common problem for the power industry, in particular, the nuclear power industry.

The erosion (wear) rate is affected by a large number of parameters: the mechanical properties of the two-phase flow particles (material properties, particle size and shape); the properties of the wall material subjected to wear; gas properties; two-phase (gas-particles) flow velocity; particle concentration in the flow; the angle of particle impact on the wall surface; flow channel geometry. For the spiral convective gas cooler (GC) used in a Shell gasification-based combined cycle unit [1], three parameters from the above list are variables: flow velocity, particle concentration, and particle impact angle. According to literature data, the dependence of erosion on the flow velocity (u) is a power function with an exponent of 2-4 [2], the dependence of erosion on the particle concentration (γ) is linear, and its dependence on the particle incidence angle (α) has a maximum at $\alpha = 25–45°$ [3, 4].

The GC tubes are made of Inconel nickel alloy, and its erosion resistance is 10–12% higher than that of stainless steel [5]. Therefore, for conservative results in our case, we chose the original empirical constants according to [6]; for the steel surface, they are $K_1 = 1.51\cdot10^{-6}$, $K_{12} = 0.296077$, and $K_3 = 5.0\cdot10^{-12}$; $\alpha_0 = 25°$.

The simulation of erosion wear and design optimization of the spiral convective gas cooler (GC) was performed using CFD software; turbulence was taken into account by means of a realizable k-ε model. The computational domain was a sector equal to 1/8 of the total cross section of the GC. The outer diameter of the tubes is 44 mm; the relative pitch between the coils of the tube spiral in the horizontal

Content from this work may be used under the terms of the Creative Commons Attribution 3.0 licence. Any further distribution of this work must maintain attribution to the author(s) and the title of the work, journal citation and DOI.
Published under licence by IOP Publishing Ltd

and vertical directions is 88 mm; the distance from the axis of the outer spiral tubes to the body is 44 mm. The computational grid was hexahedral. The parameters of the syngas are presented in Table 1 [1]. The particle diameter is varied from 2 to 100 μm, and the density is 2800 kg/m^3. It is assumed that the particles entering the computational domain have a uniform distribution over the cross section.

Table 1. Initial parameters

Flow	Parameters	Unit of measurement	Value
Syngas	mass flow rate	kg/s	113.1
	temperature	°C	750
	pressure (abs.)	bar	42
	CO, CO_2, H_2, N_2	% (vol.)	59.2; 5.1; 28.5; 7.2
Ash	mass flow rate	kg/s	1.83
	density	kg/m^3	2800
	number of particles	pcs	112414

Erosion was calculated using the Tabakoff-Grant model developed to calculate erosion of steel surfaces by ash particles [6]:

$$E = K_1 f(\alpha_1) V_1^2 \cos^2 \alpha_1 \left[1 - R_T^2\right] + f(V_{IN}) \quad (1)$$

$$R_T = 1 - 0.005249 V_1 \sin \alpha_1 \quad (2)$$

$$f(\alpha_1) = \left[1 + C_K K_{12} \sin\left(\frac{90}{\alpha_0}\alpha_1\right)\right]^2 \quad (3)$$

$$f(V_{IN}) = K_3 (V_1 \sin \alpha_1)^4, \quad (4)$$

where E is the erosion defined as the ratio of the mass loss of the obstacle to the mass of the impinging particles, mg/g; V_1 is the particle velocity, m/s; α_1 is the angle of particle incidence in degrees; α_0 is the angle of particle incidence at which the maximum erosion rate is achieved; C_K is the recovery factor for impact in the tangential direction, $C_K = 1$ at $\alpha_1 \leq 3\alpha_0$ and $C_K = 0$ at $\alpha_1 > 3\alpha_0$; K_1, K_{12}, and K_3 are empirical constants.

In the simulation of the standard GC design with a flat baffle, a high level of erosion of the metal surface of the S2.2 tube element (the second tube spiral - the second turn) was obtained. The erosion on it was $3 \cdot 10^8$ kg/(m$^2 \cdot$s) (Figure 1), and on the S2.1 and S2.3 tube elements, the maximum erosion reached $0.2 \cdot 10^8$ and $0.6 \cdot 10^8$ kg/(m$^2 \cdot$s), respectively (Figure 2). On the rest (most) of the heat-exchange surface of the GC, the erosion does not exceed $0.2 \cdot 10^8$ kg/(m$^2 \cdot$s), i.e., an order of magnitude lower.

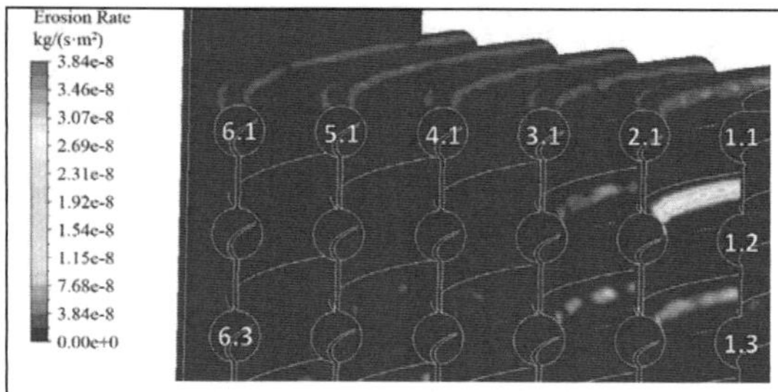

Figure 1. Erosion of GC heating surfaces.

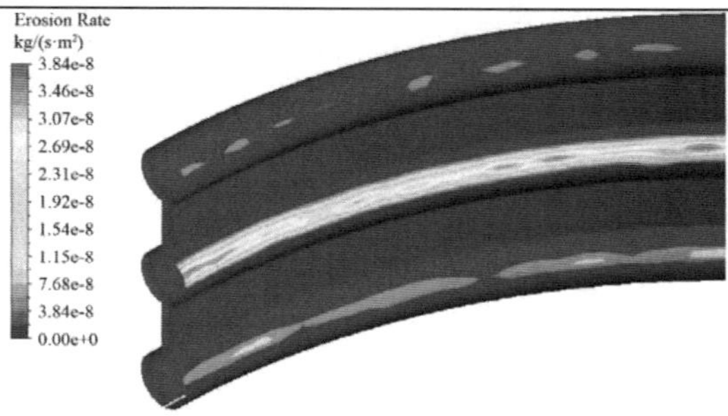

Figure 2. Erosion of the second tube spiral.

This is due to the fact that the ash particles directed to the flat baffle are mainly displaced into the nearest annular channel S1-S2 (between the 1st and 2nd tube spirals). As a result, the trajectories of particles at the inlet of the channel deviate noticeably from the vertical and the main flow of these particles collides with the membrane surface and the S2.2 tube element (Figure 3). Let us examine how this affects the change in the three variable parameters u, γ, and α.

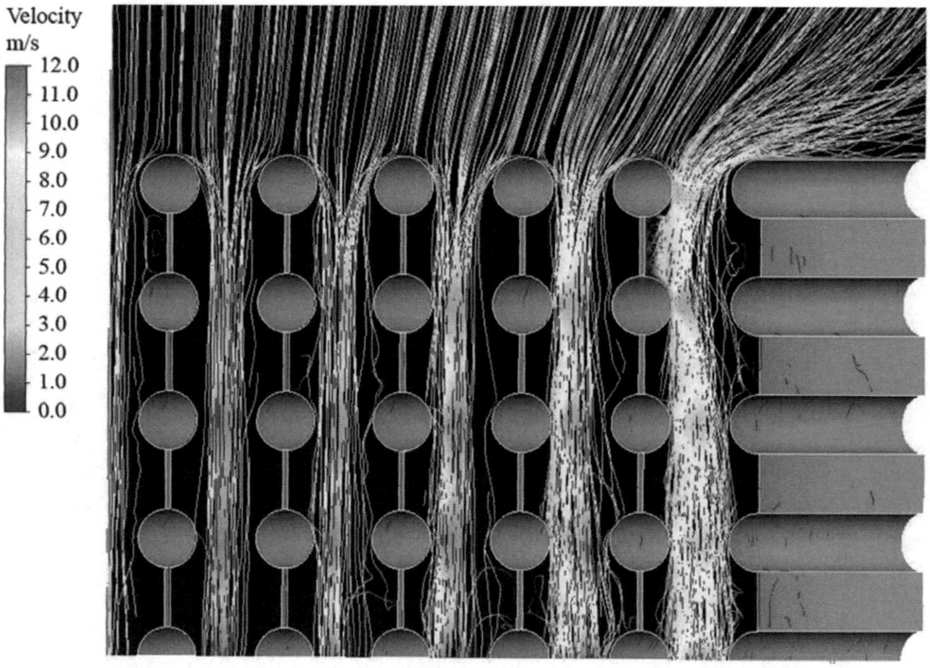

Figure 3. Particle trajectories.

As can be seen from Figure 4, the maximum particle concentration (γ) is observed at the junction of the membrane and the S2.2 tube element and reaches 1.7 kg/m^3, while the wear varies from 0 to $3.8 \cdot 10^8$ kg/(m^2·s). The maximum wear is observed in the other area, on the surface of the S2.2 tube element at φ = 55-60° (Figures 2 and 5b) and reaches $3.8 \cdot 10^8$ kg/(m^2·s).

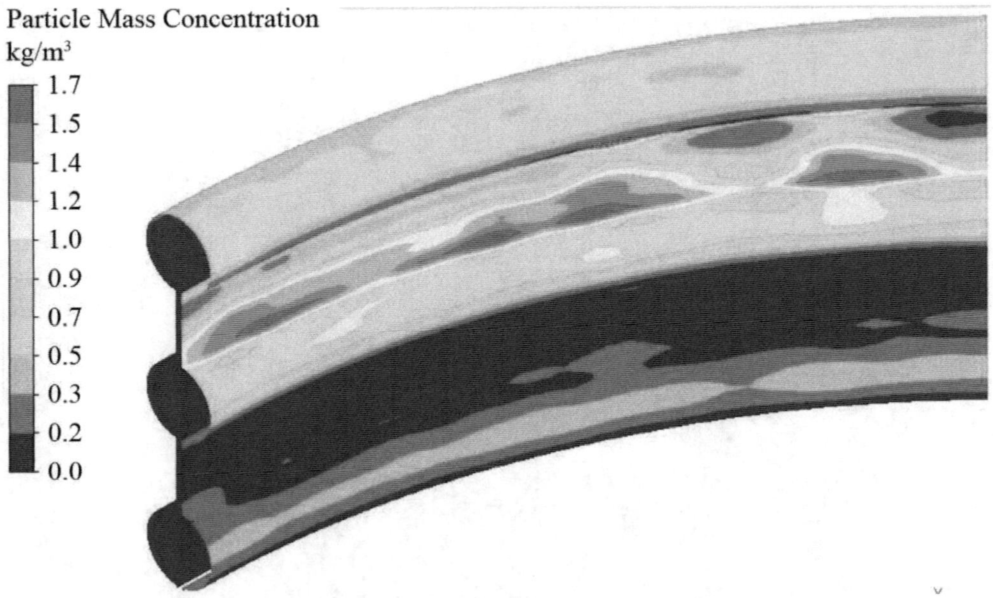

Figure 4. Concentration of particles on the surfaces of the second tube spiral.

Thus, for the given set of structural characteristics of the GC and its operating parameters, the concentration of particles is not a factor that determines the maximum erosion. Consequently, there is no need to equalize the concentration of particles over the GC cross section using, e.g. additional elements that redirect part of the particle flow.

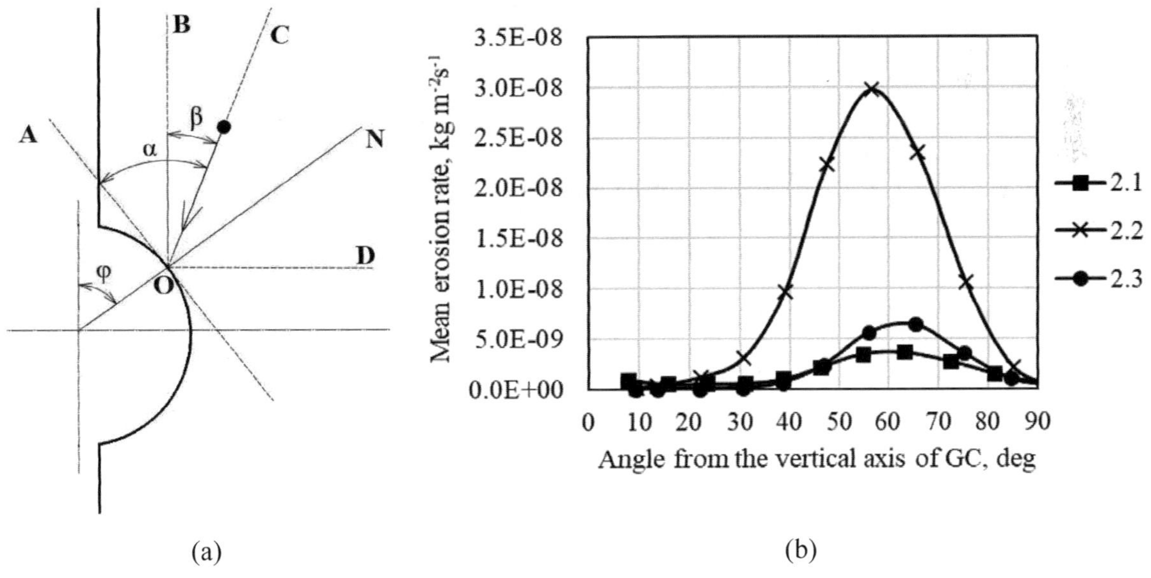

Figure 5. Intensity of erosion along the profile of the second tube spiral.

As can be seen from Figure 6, in all channels (between the S1-S6 tube spirals), the gas velocity changes along the vertical coordinate; in the narrowest sections (between the tube elements), it reaches ~11 m/s, and in the widest sections (between membranes) ~8 m/s. In contrast to the gas flow, the trajectories of particles in the channels between the S2-S6 tube spirals are practically vertical (Figure 3)

Journal of Physics: Conference Series **2150** (2022) 012017 doi:10.1088/1742-6596/2150/1/012017

and the particle velocity practically does not change with the vertical coordinate; in this case interaction of particles with the surface of the tube elements occurs only in the narrowest sections and along the tangent to the surface.

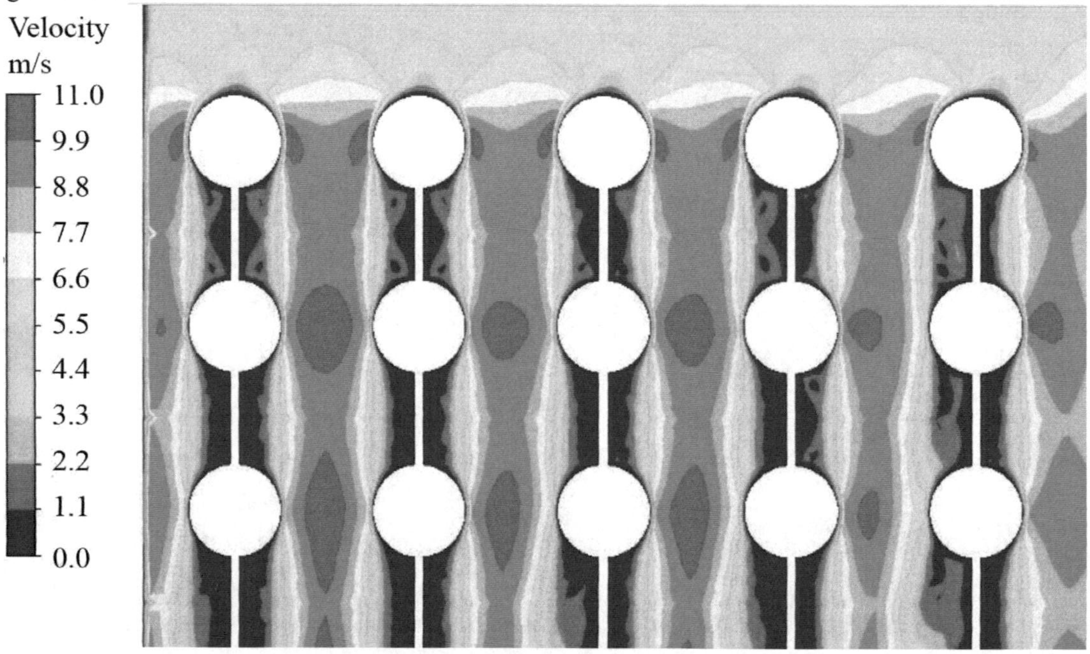

Figure 6. Gas velocity.

A different picture of particle motion is observed in the S1-S2 channel, in the initial part of the channel (from the first to the second turn), where, due to collisions with the baffle and then with the membrane and the S2.2 tube element, the particle velocity is lower than in the other channels. However, in the S2-S6 channels (with higher particle velocity), the wear is minimal — less than $0.2 \cdot 10^8$ kg/(m^2·s), and it can therefore be concluded that for a given set of structural characteristics of the GC and its operating parameters, the particle velocity is not a factor that determined the maximum erosion.

The third parameter highlighted in the analysis of the effect on the erosion rate is the angle of impact of particles on the surface α. If we assume that particles collide with the S2.2 tube element moving vertically, then for the cylindrical surface, $\alpha = 90 - \varphi$. According to the simulation results, the maximum wear is obtained at $\varphi = 55°$ (Figure 5), i.e., $\alpha = 35°$, which agrees with the data of [3, 4], where at flow velocities of up to 30 m/s, the experimental dependences of erosion on the particle impact angle have an extreme character with a maximum at angles 25–45°.

Thus, based on the results of simulation of a standard GC design with a flat baffle used for a Shell gasification-based combined cycle unit , it can be concluded that the particle impact angle α is the main factor determining the erosion maximum.

Consequently, to reduce erosion, it is necessary to install a structural element instead of the flat baffle, which allows aligning the trajectory of the ash particle flow at the inlet to the S1-S2 channel. Modeling was conducted for three shapes of the baffle — cylindrical, conical, and hemispherical. A hemispherical baffle was found to be optimal (Figure 7).

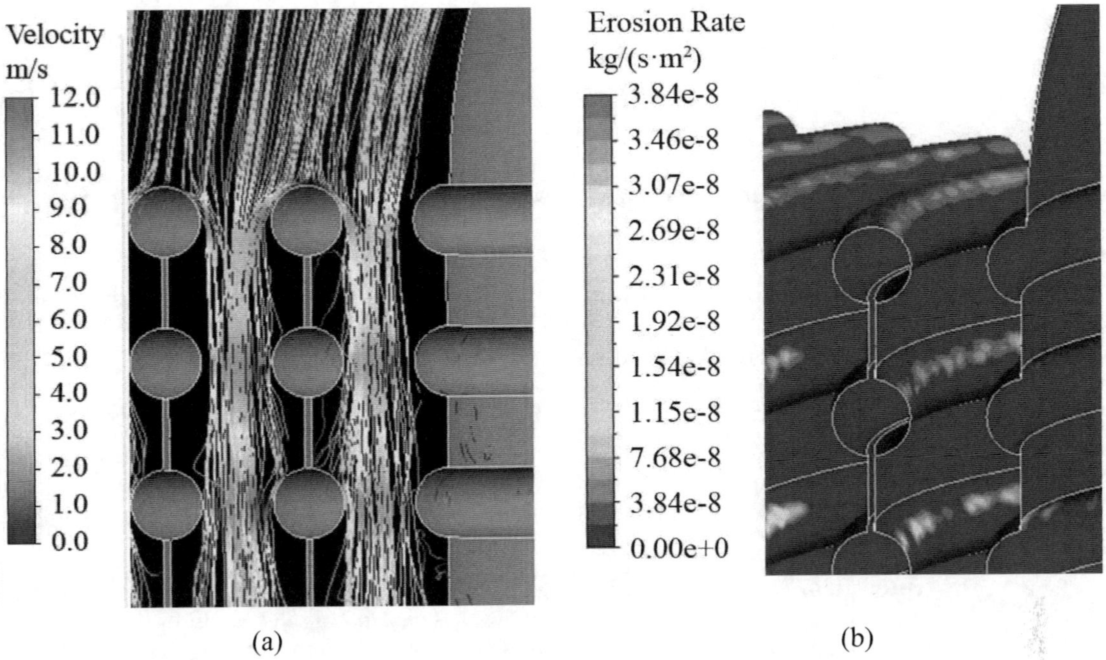

Figure 7. Particle trajectories and erosion with the installed hemispherical baffle.

As can be seen from Figure 7, the use of a hemispherical baffle makes it possible align the trajectory of the ash particle flow at the inlet to the channels and reduce the maximum erosion to 10^8 kg/(m^2·s), i.e., by a factor of almost 4 times compared to the standard geometry of the baffle.

References

[1] Park S, Ye I.–S, Oh J, Ryu C and Koo J H 2014 Gas and particle flow characteristics in the gas-reversing chamber of a syngas cooler for a 300 MWe IGCC process *Appl. Therm. Eng.* **70** pp 388–96

[2] Parsi M, Najmi K, Najafifard F, Hassani S, McLaury B S and Shirazi D A 2014 A comprehensive review of solid particle erosion modeling for oil and gas wells and pipelines applications *J. Nat. Gas Sci. Eng.* **21** pp 850–73

[3] Mbabazi J G, Sheer T J and Shandu R 2004 A model to predict erosion on mild steel surfaces impacted by boiler fly ash particles *Wear* **257** pp 612-24

[4] Raask E 1985 Mineral impurities in coal combustion: behavior, problems and remedial measures (Washington: Hemisphere Publishing Corporation) 484 p

[5] Levin B F, Vecchio K S, DuPont J N and Marder A 1999 Modeling solid-particle erosion of ductile alloys *Metall. and Mater. Trans.* **30** pp 1763–74

[6] Tabakoff W, Kotwal R and Hamed A 1979 Erosion study of different materials affected by coal ash particles *Wear* **52** pp 161–73

Control of ship tank heating systems to increase their energy efficiency

M N Syraleva and I V Kudinovich

Krylov State Research Center, Moskovskoe shosse 44, St. Petersburg, 196158 Russian Federation

syraleva_masha@mail.ru

Abstract. An important problem in the operation of icebreakers and Arctic ships is to prevent excessive icing of ballast tanks above the waterline. Uncontrolled ice formation can damage ballast systems and lead to malfunctions of ballast systems during cargo operations. This paper presents an analysis of the dynamics of water temperature in a ballast tank when the heating system is turned on and off. Recommendations on the control algorithm for the heating system that provides maximum energy efficiency are proposed.

1. Introduction

An important task in designing Arctic vessels is to choose the optimal way to control the internal icing of ballast tanks. Excessive ice formation in ballast tanks can be prevented by various methods, of which the most important are water heating and enhancing water circulation in the tank (air and steam bubbling, forced water circulation).

Previously [1], a comparative analysis of the energy efficiency (the energy efficiency criterion was the total heat flux on the ice surface providing a given ice thickness in relation to the power required for the operation of the system) of recuperative heat exchange, air bubbling, and forced circulation of water has been made, showing that the most energy efficient anti-icing systems are ballast water circulation and air bubbling systems. However, these systems have several disadvantages. A disadvantage of ballast water circulation systems is the complex design of the circulation circuit and the need to use a high-power pump. Air bubbling systems are easy to operate, but they have limitations when operating ships at very low outside air temperatures.

Ballast water heating is the least energy efficient compared to the other heating methods; however, it remains the most reliable way to prevent icing of ballast tanks under any extreme operating conditions. Based on the results of computational studies [2], the following recommendations have been developed to improve the energy efficiency of the heating system for ship's tanks:

- us of a layer of ice growing on the inner surface of the tank as a means of thermal insulation to ensure a decrease in heat loss from the tank to the environment;
- placing the coils as close to the structural waterline as allowed by their mounting options in order to reduce parasitic losses by reducing the overheating of seawater relative to the freezing point in ballast tanks below the waterline.

The normative point method [3] currently used in the design of heating systems for ship ballast tanks does not allow a correct calculation of the thermal regime of ballast tanks of complex geometric shapes. For the thermal design of such tanks, the use of CFD codes is required. However, modeling ice

Content from this work may be used under the terms of the Creative Commons Attribution 3.0 licence. Any further distribution of this work must maintain attribution to the author(s) and the title of the work, journal citation and DOI.
Published under licence by IOP Publishing Ltd

accretion and melting by means of CFD codes is a very resource-intensive and time-consuming process which is impossible at the ship design stage.
In this regard, we analyzed the dynamics of water temperature in a ballast tank when the ballast tank heating system is turned on and off.
The objectives of this study were:

- to develop a mathematical model for the temperature dynamics of ballast tanks using different control algorithms (differing in the operating time of the system and heating power);
- to compare different heating options: regular (periodic heating with power determined by the method [3]), heating with constant power, parameter control (power regulation depending on icing);
- to formulate recommendations on the heating system control algorithm that provides maximum energy efficiency.

2. Mathematical model for the dynamics of the parameters of ballast tanks for various control parameters

The method for calculating the unsteady heating of ballast tanks is based on the following assumptions.
1. A three-point model is used (three volumes along the height of the tank, see Figure 1).
2. It is assumed that ice may be present on the entire inner surface of the wall adjacent to the outside cold air and on the upper free water surface (since the maximum filling of ballast tanks is 95% of the initial volume).

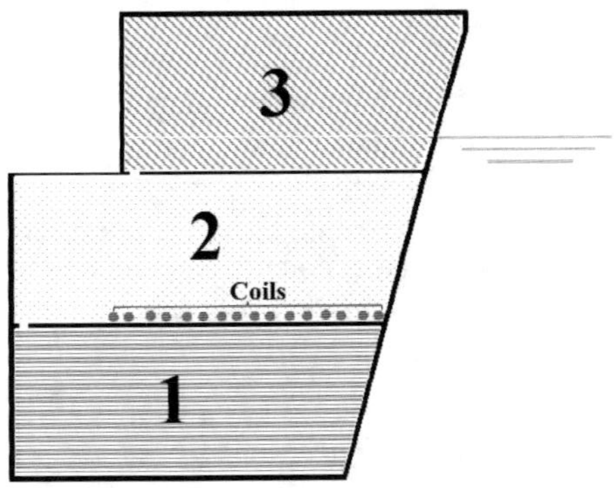

Figure 1. Schematic of the computational domain of the ballast tank

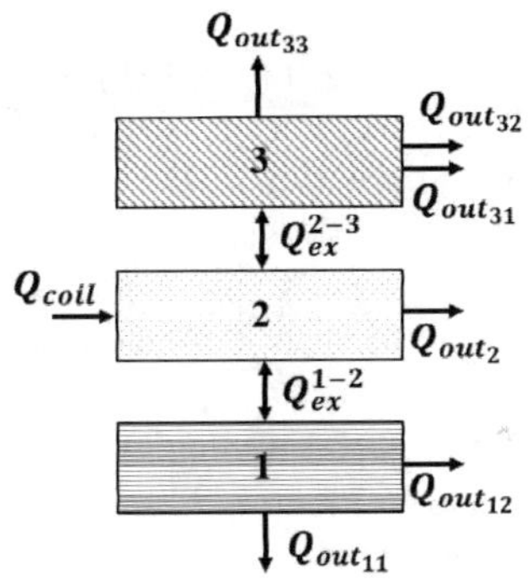

Figure 2. Schematic of the computational model of the ballast tank heating system

The dynamic temperature state and icing are given by the system of equations

$$\begin{cases} \dfrac{d\delta_{ice_{32}}(t)}{dt} = \dfrac{1}{\rho_{ice} r_{ice}}\left(q_{out_{32}}(t) - q_{out_{ice_{32}}}(t)\right); \\ \dfrac{d\delta_{ice_{33}}(t)}{dt} = \dfrac{1}{\rho_{ice} r_{ice}}\left(q_{out_{33}}(t) - q_{out_{ice_{33}}}(t)\right); \\ \dfrac{dT_1(t)}{dt} = \dfrac{1}{M_1 c_p}\left(Q_{ex}^{1-2} - q_{out_{11}}(t) S_{11} - q_{out_{12}}(t) S_{12}\right); \\ \dfrac{dT_2(t)}{dt} = \dfrac{1}{M_2 c_p}\left(Q_{coil}(t) - Q_{ex}^{2-3} - Q_{ex}^{1-2}(\Delta T) - q_{out_2}(t) S_2\right); \\ \dfrac{dT_3(t)}{dt} = \dfrac{1}{M_3 c_p}\left(Q_{ex}^{2-3} - q_{out_{31}}(t) S_{31} - q_{out_{32}}(t) S_{32} - q_{out_{33}}(t) S_{33}\right), \end{cases}$$

where $\delta_{ice_{32}}, \delta_{ice_{33}}$- are the ice thicknesses, t is time, ρ_{ice} is the ice density, r_{ice} is the heat of melting of ice, $q_{out\,i}$ is the heat flux from the wall to outside, M_i is the mass of water in the volume, c_p is the heat capacity of ice, and Q_{coil} is the heat supply.
The heat fluxes between the volumes (Q_{ex}^{i-j}) and into the outboard (Q_{out}^{i}) depend on the temperature difference between the volumes and are shown in Figure 2.
The ice thickness is determined from the condition

$$q_{ice} = q_{out}.$$

The heat flux from the water to the ice surface is

$$q_{ice} = \alpha(T - T_{ice}),.$$

where α is the heat transfer coefficient, T is the water temperature in the tank.
The heat flux from the ice surface to the outside is

$$q_{out} = \frac{T_{ice} - T_{out}}{R},$$

where $R = \frac{1}{\alpha_{out}} + \frac{\delta_{ice}}{\lambda_{ice}} + \frac{\delta_{wall}}{\lambda_{steel}}$ is the thermal resistance, δ_{wall} is the wall thickness, λ_{ice} is the thermal conductivity of sea ice, and λ_{steel} is the thermal conductivity of steel.
The values of the heat transfer coefficients for the inner surfaces are taken in accordance with recommendations [4]. The thermal resistance of partitions is $R = \frac{1}{\alpha_{vertical}} + \frac{1}{\alpha_{horisontal}} + \frac{\delta_{wall}}{\lambda_{steel}}$.

3. Simulation of the heating of a typical feed ballast tank

In this section, we consider the icing dynamics for the heating ballast tank of an ice-class vessel (the geometry of the computational domain is shown in Figure 1). The calculations are performed for a typical ballast tank size (b=0.2 m, L=7 m, H =0.7m).
The power of heating system is chosen according to the standard [3]. The initial water temperature in the ballast tank is equal to the seawater temperature (-2°C). The atmosphere temperature is -30°C.
The boundary conditions are

- the adiabatic wall on the plane surface inside the vessel;
- the ship side surface is divided into two parts: above the water line (WL) and on the deck, boundary conditions of the first kind are applied – the temperature is equal to the sea water freezing temperature T_{ice}; below the WL (underwater part), boundary conditions of the third kind are applied with outboard water heat transfer coefficient α_{out} and temperature T_{out};
- on the surface of the heating element (coil), a boundary condition of the second kind is applied – heat flux Q_{coil}.

4. Analysis of various control algorithms for ballast tank heating systems

Three different operating modes of the heating system are considered.

1. Regular mode: heating is switched on and off every 4 hours, and the system is designed to heat water from -2°C to 4°C [3].
2. Continuous heating with a power that is only 2% of the value required by [3].
3. Heating with power depending on the icing of the inner surface.

The main results are presented in Table 1 and Figure 3.

Table 1. Comparison of the three operating modes of the heating system

Operating mode	Maximum ice thickness $\delta_{ice_{max}}$, m	Maximum temperature, T_2, °C	Amount of heat supplied in 7 days, J
1	0.005	72.0	$2.2 \cdot 10^{10}$
2	0.1	0.1	$1.3 \cdot 10^{9}$
3	0.11	0.1	$1.2 \cdot 10^{9}$

When using the recommended [3] periodic heating mode of ballast tanks (period of 4 hours), there is significant overheating of ballast water, which leads to an increase in heat loss and a decrease in the energy efficiency of the heating system.

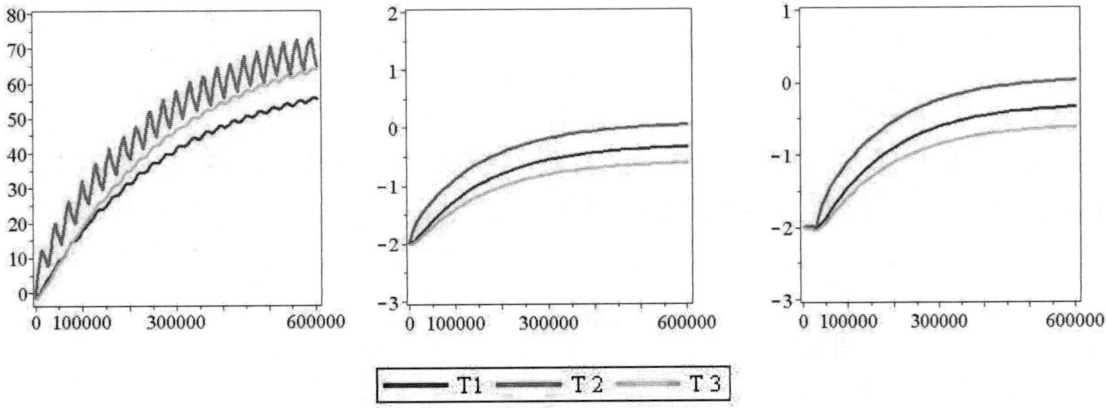

Figure 3. Dependence of temperature in the volumes on time for three operating modes of the heating system

5. Conclusions

The results of the study show that the regular operation mode of heating ballast tanks [3] is not optimal in terms of energy efficiency.

To improve energy efficiency, it is possible to recommend the operating mode of the ballast tank heating system for Arctic ships with a constant thermal power at the level of 2% of the heating system rated power calculated in accordance with the recommendations [3]. This allows maintaining a constant thickness (maximum allowable) ice on the inner surface of the tanks.

References

[1] Comparative analysis of energy efficiency of ballast tank anti-freezing systems, Journal of Physics: Conference Series, doi:10.1088/1742-6596/1683/2/022023

[2] Bagaev D V, Kudinovich I V, Syraleva M N 2019 Improving the energy efficiency of ship tank heating systems (All-Russian Scientific Conference with International Participation Thermophysics and Power Engineering in Academic Centers, TPEAC-2019, Saint-Petersburg, Russia), pp 44-45

[3] State Standard RD5R.113-85 1985 System for heating liquids in ship tanks. Design rules Moscow, p 172

[4] Kudinovich I V, Syraleva M N 2013 Study of icing-up in the fresh water tanks of the semi-submersible drilling rigs operated in northern seas (International Conference on Computer Applications in Shipbuildings, Busan, Korea), Vol. III, pp 27-32

Displacement of oil by SiO_2-nanofluid from a microporous medium: microfluidic experiments

A I Pryazhnikov[1], A V Minakov[1,2], M I Pryazhnikov[1,2], V A Zhigarev[1]and I V Nemtsev[1,3]

[1] Siberian Federal University, 79 Svobodny pr., Krasnoyarsk, 660041, Russia
[2] Kutateladze Institute of Thermophysics SB RAS, 1 Acad. Lavrentyev pr., Novosibirsk, 630090, Russia
3 Krasnoyarsk Regional Center of Research Equipment of Federal Research Center «Krasnoyarsk Science Center SB RAS», 50 Akademgorodok st., Krasnoyarsk, 660036, Russia

arrivent@yandex.ru

Abstract. In this work, the process of displacing oil from a microfluidic chip that simulates a porous medium is studied. Experimental photographs of the process of oil displacement by water and SiO_2-nanofluid are presented. It is shown that the use of nanofluid increases the oil displacement efficiency by 16%.

1. Introduction

Currently, methods for improving oil recovery are being actively investigated. One of the most relevant and promising methods for improving oil recovery is nanofluid flooding [1-4]. A nanofluid means a dispersed system consisting of a carrier liquid (for example, water) and nanoparticles [5].

Microfluidic studies using microporous rock models are becoming an alternative to conventional core flooding. [6-7]. Microfluidics (μικρός (micros) - small, fluidis - fluid) is an interdisciplinary science that describes the behavior of small (volume of the order of micro- and nanoliter) flows of liquids. The essential advantages of microfluidic studies are excellent visualization of the processes of oil displacement from the porous medium model, simple cleaning and the possibility of multiple use of the microfluidic chip.

This article presents the results of an experimental study of oil displacement by SiO_2 nanofluid from a microporous medium. The dynamics of the displacement process is shown. The study was carried out using the "lab-on-chip" microfluidic technology. It is found that the use of nanofluid improves the oil displacement efficiency.

2. Experimental materials and methods

This section describes the fluids used, microporous media, experimental setup, and procedure.

2.1. *Properties of used fluids*

In the work, crude light oil with a density of 856 kg m^{-3} and a viscosity of 24.6 mPa s was used. Distilled water and a suspension of nanoparticles of silicon dioxide were used as displacement fluids. The mass concentration of nanoparticles was 2%. The nanofluid was obtained by diluting a Ludox TM-50

Content from this work may be used under the terms of the Creative Commons Attribution 3.0 licence. Any further distribution of this work must maintain attribution to the author(s) and the title of the work, journal citation and DOI.
Published under licence by IOP Publishing Ltd

suspension (Sigma-Aldrich) with distilled water. The viscosity coefficient of the nanofluid, measured by the method in [8], was 0.939 mPa s. The nanoparticles had a spherical shape with the primary particle size of 22 nm. An image of SiO_2 nanoparticles obtained by a Hitachi S-5500 ultra-high resolution scanning electron microscope is shown in figure 1.

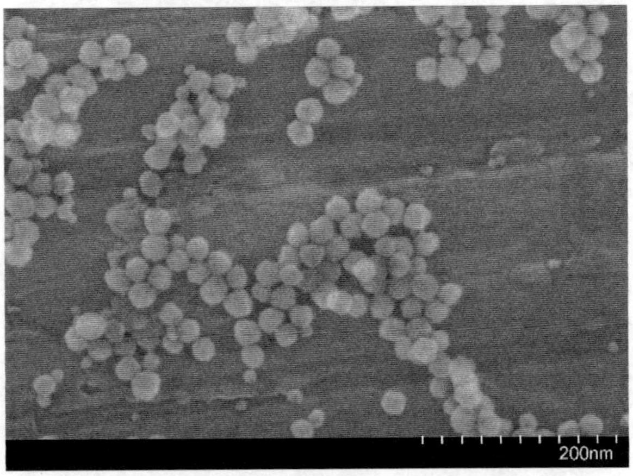

Figure 1. SEM photo of SiO_2 nanoparticles.

The particle size distribution in liquid was obtained using an acoustic and electroacoustic spectrometer DT1202 (see figure 2). Suspension of nanoparticles of silicon oxide had a high stability (zeta potential is -31 mV). The average hydrodynamic particle size was 30 nm [9].

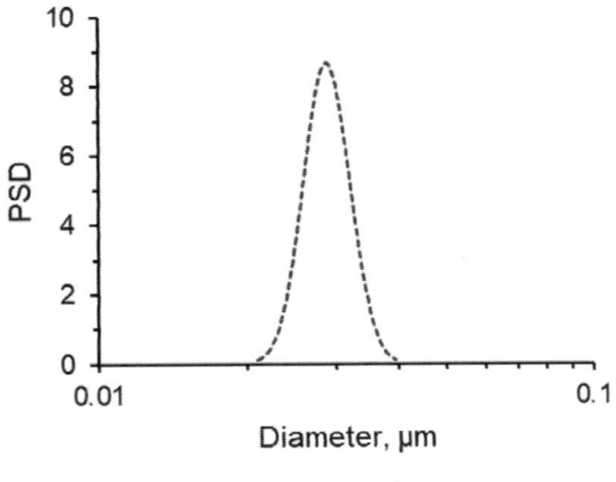

Figure 2. Particle size distribution of nanoparticles in liquid.

2.2. Microfluidic chip

In this work, a microfluidic chip (Dolomite: 3200284) is used to simulate the complex porous structure of the rock. The chip is made of glass. The porous area is 10×60 mm. The chip has one inlet and outlet. The porous area is formed by repeating (150 times) a 2×2 mm square (see Figure 3). The channel arrangement in the square is an 8×8 channel grid. Channels in the mesh have constrictions or "pores" that are randomly distributed. The mesh contains 38 pores with Ø63 µm, 40 pores with Ø85 µm and 50 direct channels (Ø100×110 µm). The pore volume is 38 µl.

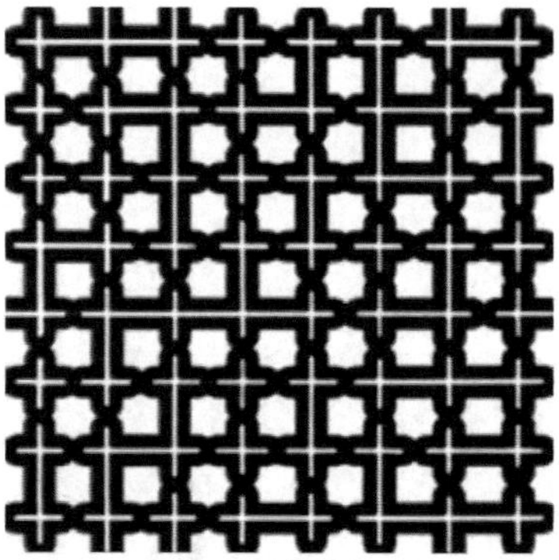

Figure 3. Microscope photo of a microfluidic chip. Magnification 10x.

2.3. Experimental setup and procedure

The displacement fluid flow is controlled using an Elveflow OB1 MK3+ microfluidic pressure controller. The controller is equipped with a pressure channel from 0 to 8 bar, the accuracy of pressure maintenance is 100 Pa. The controller requires an external pressure source to operate. The compressed air from the pressure controller enters the sealed reservoir with the investigated displacing liquid (water or nanofluid). The microfluidic chip is connected to the reservoir with a 1/16 "OD PTFE tubing. The microfluidic chip is placed horizontally on a glass slide. The high-speed camera is located at the top above the chip, and the source is at the bottom. An MFS3 flow sensor, operating in the range 2.4 to 80 μl min^{-1} with an accuracy of ± 5% of the measured value, is used.

The empty chip was first filled completely with oil, and then the process of flooding with fluid displacement took place at a fixed flow rate. Two to three pore volumes (PV) were pumped. The picture of the waterflooding process was recorded by a high-speed camera. After each experiment, the microfluidic chip was thoroughly sequentially washed with dichloroethane, isopropanol, distilled water, and air.

3. Results and discussions

First, an experimental microfluidic waterflooding was carried out. The microfluidic chip, completely filled with oil, was supplied with water at a volumetric flow rate of 45 μl min-1. This corresponded to the capillary number Ca (a dimensionless parameter that determines the ratio of viscous forces to interfacial tension forces) of $1.6 \cdot 10^{-3}$. The dynamics of the process of oil displacement by water is presented in figure 4.

A special application was developed to calculate the oil displacement efficiency based on photographs of the distribution of oil and the displacing liquid in the pore space of a microfluidic chip. A set of images was obtained by converting video recordings using the free FFmpeg library. To develop the application, the BlackBox Component Builder component framework (Oberon microsystems, Switzerland) and the FreeImage library were used. The HSV (Hue-Saturation-Value) color model was used when analyzing the proportion of oil in a microfluidic chip. This application could detect the pixels occupied by oil, and thus water and oil saturation at a given time was determined.

Figure 4 shows the photos of the process of oil displacement by water at different points in time. The front of the water flow moves in the form of jets. Such jets quickly break through to the outlet, in about 10 s. After all the jets reach the outlet, the flow stabilizes. The displacement of oil stops.

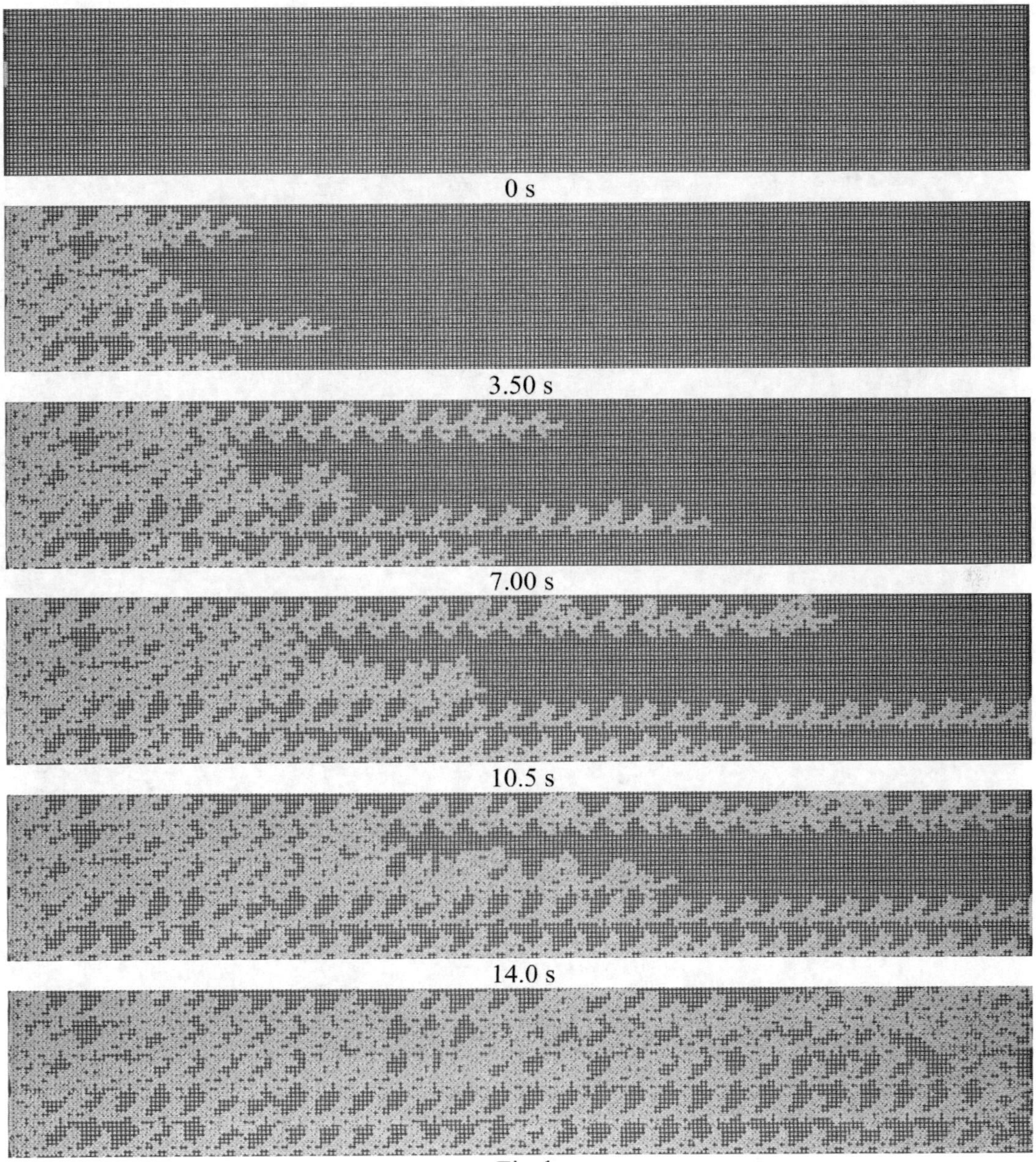

Figure 4. Photos of the process of displacing oil with water from a microfluidic chip.

The displacement of oil by nanofluid from a microporous medium occurs in a significantly different way. As it is shown in figure 5, the movement of nanofluid occurs with a more uniform front throughout the entire volume of the microfluidic chip. The nanofluid breaks through to the outlet from the microporous model later than water. The breakthrough time of the nanofluid before leaving the microfluidic chip is 15 s. Therefore, more oil is displaced with the help of nanofluid as compared to water.

The final distributions of oil fraction during displacement by water and nanofluid (see the last experimental flow patterns in Figures 4-5) differ significantly. It is clearly seen that when displaced by water, large areas filled with oil remain, which are held by capillary forces. When using a nanofluid as a displacing fluid, there are much fewer such regions.

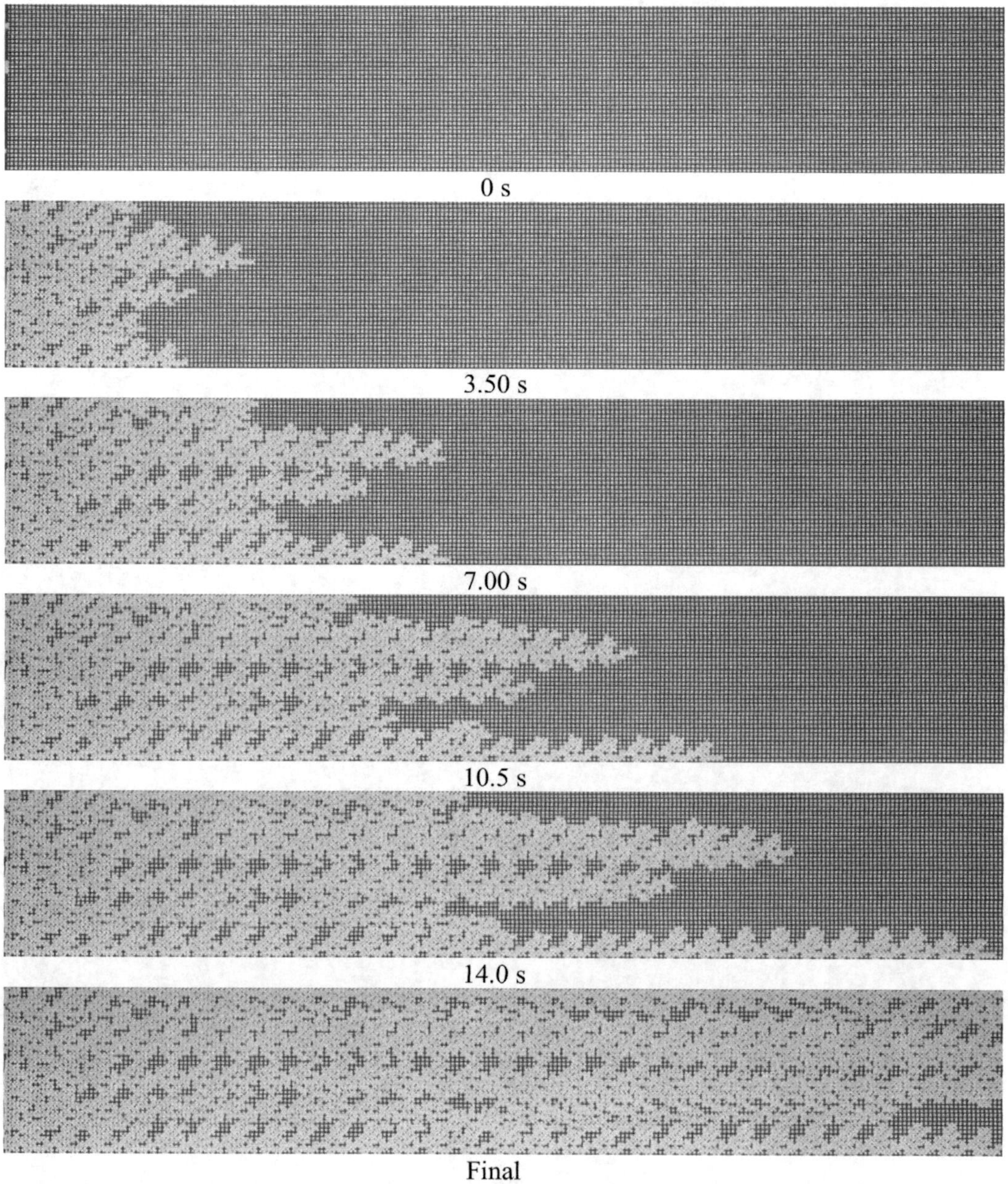

Figure 5. Photos of the process of displacing oil from a microfluidic chip by nanofluid.

The dependence of the oil displacement factor on time is shown in figure 6. Figure 6 shows that at the initial moment, the oil recovery factor (ORF) increases linearly (in proportion to the flow rate of the displacing fluid). After the displacement fluid reaches the outlet, the growth of ORF is slowed down. After pumping one pore volume, the ORF value does not change. It was found that ORF during water flooding was 60%, and it was 76% during nanofluid flooding.

There are many factors influencing the oil displacement efficiency. These factors include viscosity, density, interfacial tension, and wettability. However, as it is noted earlier [10], the main factor is wettability. Nanofluid affects the wettability, contributing to the washing out of capillary retained oil [11].

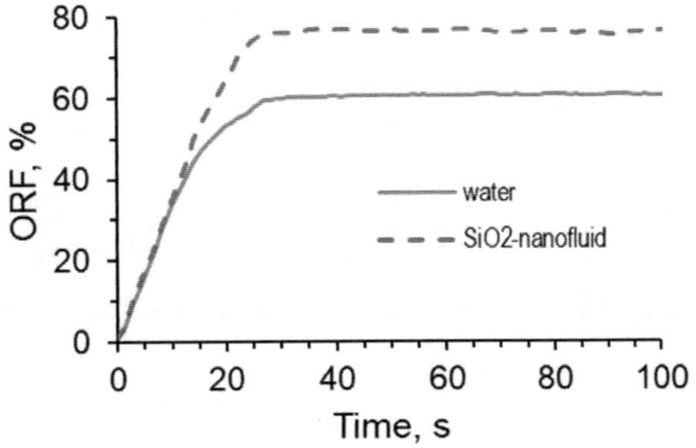

Figure 6. Dynamics of change in the oil recovery factor in time.

4. Conclusions

In this work, an experimental study of the process of displacing oil with water and 2% SiO_2-nanofluid from a microporous medium was carried out. The study was carried out using the "laboratory on a chip" technology. Microfluidic chip with microporous medium simulating rock was used.

As a result, the dynamics of the oil displacement process versus the time of the displacing fluid (water and nanofluid) was obtained. It is shown that 2% -nanofluid increases the oil recovery factor by 16% as compared to water.

Acknowledgments

This paper is partially financed by the Ministry of Science and Higher Education of the Russia (no. FSRZ-2020-0012).

We are grateful to the Krasnoyarsk Regional Shared Research Center (Krasnoyarsk Scientific Center, Siberian Branch, Russian Academy of Sciences) for taking characterization of nanoparticles.

References

[1] Nasr M S, Esmaeilnezhad E and Choi H J 2021 *J. Taiwan Inst. Chem. Eng.* **122** 241-59
[2] Minakov A V, Guzei D V, Pryazhnikov M I, Filimonov S A and Voronenkova Y O 2021 *Petrol. Explor. Develop.* **48**(4) 956–67
[3] Yakasai F, Jaafar M Z, Bandyopadhyay S and Agi A 2021 *J. Ind. Eng. Chem.* **93** 138-62
[4] Cao J, Chen Y, Zhang J, Wang X, Wang J, Shi C, Ning Y and Wang X 2022 *Chem. Eng. Sci.* **247** 117023
[5] Minakov A V, Rudyak V Y and Pryazhnikov M I 2018 *AIP Conf. Proc.* **2027** 030141
[6] Betancur S, Olmos C M, Pérez M, Lerner B, Franco C A, Riazi M, Gallego J, Carrasco-Marín F and Cortés F B 2020 *J. Pet. Sci. Eng.* **184** 106589
[7] Zhao X, Feng Y, Liao G, Liu W 2020 *Colloid Interface Sci.*, **578** (2020) 629–40
[8] Minakov A V, Rudyak V Y and Pryazhnikov M I 2021 *Heat Transfer Eng.* **42**(12) 1024–40
[9] Pryazhnikov M I and Minakov A V 2020 *Tech. Phys. Lett.* **46**(6) 606-9
[10] Minakov A V, Pryazhnikov M I, Zhigarev V A, Rudyak V Y, Filimonov S A 2021 *Theor. Comput. Fluid Dyn.* **35**(4) 477-93
[11] Minakov A V, Pryazhnikov M I, Suleymana Y N and Meshkova V D 2020 *Tech. Phys. Lett.* **46**(12) 1238–40

Improving the maneuverability and thermal efficiency of modern cogeneration systems based on gas turbine power plants

E M Lisin[1] and V O Kindra[2]

[1] Department of Economics in Power Engineering and Industry, National Research University "Moscow Power Engineering Institute", Moscow, 111250 Russia
[2] Department of Innovative Technologies for High-Tech Industries, National Research University "Moscow Power Engineering Institute", Moscow, 111250 Russia

lisinym@mpei.ru

Abstract. The paper is devoted to the issue of increasing the maneuverability and efficiency of modern cogeneration systems based on gas turbine power plants. Promising solutions for increasing the maneuverability of GTU-CHPP by using heat accumulators and the formation of a preheating circuit of the network water are considered. It is shown that in the non-heating period, it is possible to increase both the thermal efficiency and the generated electric power by installing a heat exchanger in front of the compressor. The calculation results show that this provides an increase of 0.4% in the net electrical efficiency by and an increase 3.3% in the annual electricity generation.

1. Introduction

The long-term centralized development of the electric power industry in Russia the high efficiency of combined production, and climate features (the thermal power consumption is two times higher than the electric power consumption) has led to the wide use of CHPPs and centralized power systems based on them in the country. These centralized power systems meet up to 90% of the urban heat and electricity demands [1, 2, 3].

During the planned economy period, CHPPs were rationally integrated into the unified energy system of the country and exhibited high performance. However, after the transition to a market economy, the position of CHPPs has deteriorated significantly due to the weak adaptability of this type of generation to market conditions. Over the past 20 years, heat supply from CHPPs decreased by a factor of more than 1.5. Despite this negative trend, CHPPs still play a major role in the energy supply to consumers. They account for about 47% of electricity generation by thermal power plants and 37% of heat supply by heat sources [4, 5, 6].

The main advantages of cogeneration of energy products at CHPPs are [6, 7]:

- low specific fuel consumption in the combined generation of electricity and heat,
- the possibility of separating conditionally variable costs (fuel savings due to district heating) between electricity and heat, depending on the situation in the power markets,
- production of electricity close to the place of its consumption,
- increased reliability as a heat source.

Content from this work may be used under the terms of the Creative Commons Attribution 3.0 licence. Any further distribution of this work must maintain attribution to the author(s) and the title of the work, journal citation and DOI.
Published under licence by IOP Publishing Ltd

At the same time, combined production at CHPPs has a number of disadvantages that do not allow its advantages to be fully realized in market conditions [6, 8]:

- low maneuverability in terms of electrical unloading of generating equipment operating in cogeneration mode due to the need to cover the heat load (impossibility of reducing the supply of electrical energy without simultaneously reducing the supply of heat),
- a small number of hours of operation of the installed capacity in heating mode due to the seasonal nature of the demand for heat power.

The electricity generated by different types of CHPPs in the heating cycle ranges from 25 to 150% of the amount of heat consumption. At the same time, the unit cost of electricity production is approximately the same, regardless of the type of power plant (Table 1). Thus, by changing the composition of power equipment and production technology (heat cycle), it is possible to produce electricity and heat in different ratios with the same efficiency. Since CHPPs are usually built, expanded, and reconstructed over decades (due to a gradual increase in the load and a change in the structure of electricity and heat production), many plants have power units of various types [5, 6, 9, 10].

Table 1. Comparative indicators of specific fuel consumption for various types of TPPs

Power plant type and operating mode		Products	Fuel utilization factor, %	Specific fuel consumption for electricity production, g / kWh
STU-CHPP	heating mode	electricity and heat	82	150
	condensing mode	electricity	35	351
CCGT-CHPP	heating mode	electricity and heat	82	150
	condensing mode	electricity	51	241
GTU-CHPP	combined production	electricity and heat	82	150
CPP		electricity	39	315
CCGT plant		electricity	55	212
GTPP		electricity	35	340

In connection with the development of distributed generation, GTU-CHPPs, which have relatively low capital costs and a high fuel utilization factor, have gained particular importance in ensuring independent power supply of regions [1, 2]. At the same time, the development of independent resource-saving regional power systems based on GTU-CHPPs necessitates an increase in their capacity and efficiency. This requires the development of new highly efficient thermal schemes.

2. Methods

Despite the possibility of significant fuel savings in the joint production of energy products at CHPPs, combined generation is being pushed out of the production structure of territorial power systems. The reason for this is the technological limitations of the production process, such as low maneuverability, a limited share of energy production based on heat consumption during the year, and a decrease in electricity generation during heat extraction.

Uneven consumption of energy products forces power plants to operate with frequent changes in operating modes, which leads to a decrease in the efficiency of the production process and a reduction in the service life and reliability of the main generating equipment. There are several ways to level the load curve of thermal power plants [11, 12]:

- use of multi-part tariffs to stimulate steady consumption of energy products by consumers;
- peak-load operation of part of the thermal power plant equipment;
- energy storage at thermal power plants and the use of additional heating circuits to heat the network water.

The use of energy storage systems is the most promising production and technological solution for leveling the load schedule and increasing the maneuverable characteristics of CHPPs. Increased maneuverability is achieved by increasing the ability of a power system to quickly respond to a change in its power, characterized by the adjustable range of operation and the rate of change of load.

Whereas the use of energy storage systems is limited by their low capacity and high cost, the use of heat accumulators is much more technological and economically feasible [13, 14]:

- significant expansion of the power regulation range (20–25% for electricity supply and 40-50% for heat supply),
- high rate of capacity changes due to short-term and fairly simple operations (start-stop of pumps, opening-closing of valves),
- relatively low capital costs for additional capacity.

The most practical solution is to use hot water storage tanks as heat accumulators in the heat supply scheme from CHPPs. At the same time, to ensure operability, it is necessary to have two accumulator tanks for separate storage of hot and cold water. During peak electrical loads, excess heating water is discharged from the cold water storage tank. Having passed through the network heaters, the excess part of the already heated network water is directed to a hot water accumulator. Increasing the heat load leads to the generation of additional electricity based on heat consumption. During the hours when the electrical load is disconnected, water from the hot heating water accumulator is supplied to the input to the network heater. The heat load of the turbine unit is reduced, and its electrical power is proportionally reduced.

It is also possible to increase the generation of electricity based on heat consumption by including a preheating circuit of network water in the thermal scheme of a power plant (Figure 1). This solution is most effective for GTU CHPPs [15, 16, 17].

Air is pumped in by a compressor, compressed to a higher pressure, and sent to the combustion chamber. Fuel (natural gas) is also fed into the combustion chamber, where it burns, providing a constant supply of heat. The compressed air mixes with the combustion products to form a high-temperature gas that enters the gas turbine, where it expands to atmospheric pressure and does useful work by generating electricity. After the turbine, the exhaust gases are utilized in a gas-water heat exchanger, where they are used to heat water in the heating system and, accordingly, to supply heat to consumers.

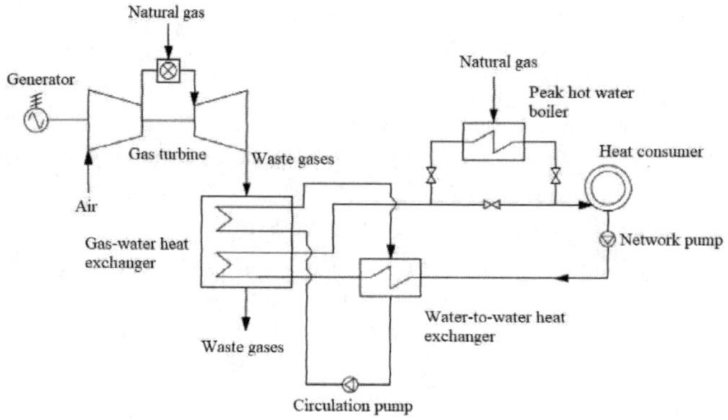

Figure 1. Diagram of a GTU-CHPP with a preheating circuit for network water

The gas-water heat exchanger consists of two heating surfaces. The first heating surface along the gas path serves to transfer heat to the coolant circulating in a closed preheating circuit of return network water. For this, a water-to-water heat exchanger is used. The heated network water is directed to the second heating surface of the gas-water heat exchanger, where it is further heated to the required temperature according to the temperature schedule of the heating network and is supplied to the consumer.

3. Results and discussion

Figures 2 and 3 show the results of performance analysis showing the advantages of using a preheating circuit for network water in a GTU-CHPP.

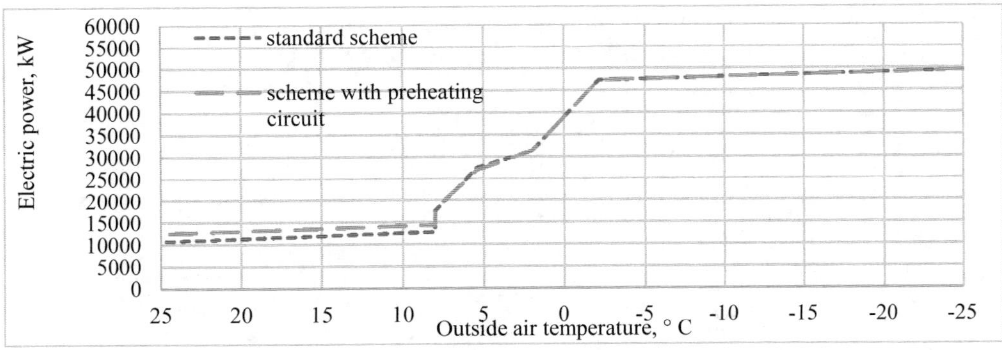

Figure 2. Increase in electric power due to the use of a preheating circuit of network water in a GTU-CHPP

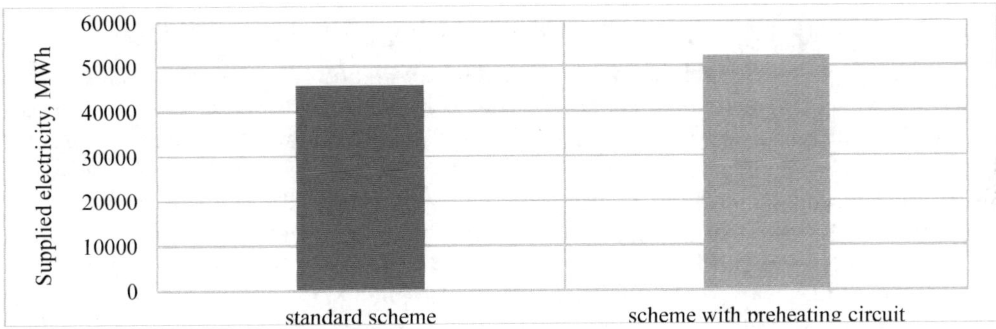

Figure 3. Increase in supplied electricity supply in summer when using a preheating circuit of network water in a GTU-CHPP

Comparison of the energy efficiency of thermal schemes of GTU-CHPP is made on the basis of the condition of covering a given heat load schedule for climatic conditions typical for the Moscow region. The ratio of the heat load in the non-heating season to the maximum heat load is assumed to be 1: 5. The main characteristics of thermal schemes are shown in Table 2.

Table 2. Main characteristics of thermal schemes of GTU-CHPP

Specifications	Standard scheme	Scheme with preheating circuit	
Number of power units, pcs	3		
Gas turbine model	Siemens SGT-400		
Net electric power of CHPP in design mode, MW	41.2		
Heat output of the CHPP in design mode, MW	83.4		
Design point of calculation of a gas-water heat exchanger, °C	-2.2		
Design point of calculation of the second in the course of gases heating surface of the gas-water heat exchanger, °C	-	2	25
Peak source of heat	Peak hot water boilers		
Heat supply control method	Uniform unloading of gas turbines and shutdown of the number of operating power units		

The advantage of using a preheating circuit for network water in a GTU-CHPP is an increase in the electrical power generation based on heat consumption, primarily in summer, when the demand for heat is quite low. A disadvantage is an increase in electricity consumption for own needs due to the need to ensure the operation of the circulating pump in the closed circuit of preliminary heating of network water. The thermal efficiency of a GTU-CHPP can be increased using the following thermal diagram presented in Figure 4.

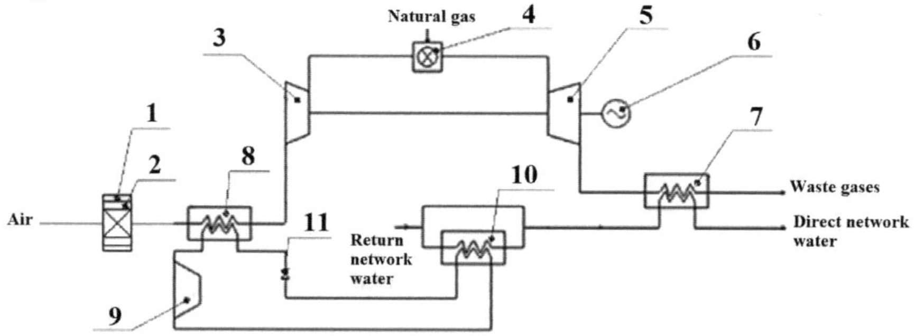

Figure 4. Schematic diagram of a highly efficient GTU-CHPP

Atmospheric air passes through an air cleaning device (1), and enters compressor (3), where it is compressed and directed to the inlet of combustion chamber (4). After the combustion of the hot mixture and the generation of useful work in gas turbine (5), hot gaseous combustion products are directed to the inlet of the hot coolant circuit of the gas-water heat exchanger (7), where they heat the heating water entering the cold circuit of the heat exchanger after the hot coolant circuit of the evaporative heat exchanger (10). At the outlet of the hot coolant circuit of the gas-water heat exchanger (7), gaseous products are discharged into the atmosphere in the form of flue gases. Network water from the outlet of the coolant circuit of the gas-water heat exchanger (7) is directed to the consumer. Electric generator (6) is used to generate a payload and electricity to power a compressor (3). Calculations for the proposed thermal scheme show that heating of atmospheric air from - 20 °C to 10 °C in the heat exchanger during the heating period allows increasing the net electrical efficiency by 7.5% compared to the standard CHPP scheme by reducing the cost of electricity for heating air in the electrical heat exchanger (2). In turn, cooling of atmospheric air in the non-heating period from 20 °C to 10 °C in the heat exchanger (7) in the non-heating per allows increasing the net electrical efficiency by 0.4% due to an increase in air consumption at the compressor inlet (3) and due to a decrease in the specific power of the compressor. This provides a 3.3% increase in annual electricity generation and a 5.8% increase in annual heat supply.

Conclusions

Currently, the construction of efficient cogeneration systems is the most important condition for solving the problems of energy and resource conservation outlined in the Energy Strategy of Russia for the period up to 2035. The cogeneration system involves the implementation of joint heat and power generation at CHPPs using steam turbine and gas turbine equipment. In connection with the development of distributed generation, GTU-CHPPs, which are distinguished by relatively low capital costs with a high fuel utilization factor, have gained particular importance in ensuring the independence of power supply to the regions.

Despite the above advantages, the development of independent resource-saving regional power systems based on GTU-CHPP requires an increase in their capacity, maneuverability, and efficiency. This necessitates the development of new highly efficient thermal schemes.

The most effective solution for increasing electricity generation in combined production for GTU-CHPPs is to use a preheating circuit for network water in the thermal scheme of the power plants. At the same time, this leads to an increase in electricity consumption for own needs.

It is possible to increase the thermal efficiency simultaneously with the supplied electric power by installing a heat exchanger in front of the compressor in the non-heating period. Calculation results show that a 0.4% increase in the net electrical efficiency is due to an increase in air consumption at the compressor inlet and a decrease in the specific power of the compressor. This provides a 3.3% increase in annual electricity generation by and a5.8% increase in annual heat supply.

Acknowledgments
This paper was supported by a grant of RF President for governmental support of young scientists PhD and DSc (MD-914.2020.6)

References

[1] Lisin E , Marishkina Y, Strielkowski W, Streimikiene D 2017 Analysis of competitiveness: energy sector and the electricity market in Russia *Economic Research-Ekonomska Istraživanja* **30** pp 1820-1828
[2] Lisin E 2017 Development and research of structural and technological modernization scenarios of power engineering in power market conditions *Journal of Physics: Conference Series* **891**(1) p 012198
[3] Volkova E A., Makarova A S, Khorshev A A, Urvantseva L V, Shul'gina V S 2010 Investigating the efficiency of development of centralized heating in Russia *Thermal engineering* **57**(14) pp 1204-1215
[4] Lisin E, Kindra V, Horvathova Z 2017 Sustainable development of regional heat supply systems in the context of the Eurasian Economic Union energy markets association *Journal of Security and Sustainability Issues* **6**(4) pp 745-760
[5] Lisin E, Sobolev A, Strielkowski W, Garanin I 2016 Thermal efficiency of cogeneration units with multi-stage reheating for Russian municipal heating systems *Energies* **9**(4) pp 269-287
[6] Filippov S P, Dil'man M D 2018 CHP plants in Russia: The necessity for technological renovation *Thermal Engineering* **65**(11) pp 775-790
[7] Korppoo A, Korobova N 2012 Modernizing residential heating in Russia: End-use practices, legal developments, and future prospects *Energy Policy* **42** pp 213-220
[8] Lisin E, Kindra V, Strielkowski W, Zlyvko O, Bartkute R 2017 Economic analysis of heat and electricity production in the decentralisation of the Russian energy sector *Transformations in Business and Economics* **16**(2) pp 75-88
[9] Lisin E, Kindra V, Egorov A, Kurdiukova G 2016 Design and feasibility study of energy-conserving thermal schemes for gas turbine-based CHP *Contemporary Engineering Sciences* **9**(32) pp 1559-1574
[10] Teplov B D, Radin Y A 2019 Improving flexibility and economic efficiency of CCGT units' operation in the conditions of the wholesale electricity market *Thermal Engineering* **66**(5) pp 323-330
[11] Zaytseva Y V 2016 Econometric modeling of electricity consumption by households as a tool for the calculating of the social consumption norm *R-Economy* **2**(2) pp 259-269
[12] Strielkowski W, Lisin E, Tvaronaviciene M 2016 Towards energy security: sustainable development of electrical energy storage *Journal of Security and Sustainability Issues* **6**(2) pp 43-52
[13] Korpela T, Kaivosoja J, Majanne Y, Laakkonen L, Nurmoranta M, Vilkko M 2017 Utilization of district heating networks to provide flexibility in CHP production *Energy Procedia* 116 pp 310-319
[14] Turunen J, Majanne Y, Vilkko M 2020 Short-term Optimization of the Operation of the CHP District Heating Plant with Heat Accumulator *IFAC-PapersOnLine* **53**(2) pp 13236-13241
[15] Lisin E, Kindra V, Egorov A, Kurdiukova G 2016 Design and feasibility study of energy-conserving thermal schemes for gas turbine-based CHP *Contemporary Engineering Sciences* **9**(32) pp 1559-1574
[16] Kindra V, Lisin E, Burov V, Egorov A 2016 Design solutions to prevent heat exchanger surface cold-end corrosion at gas turbine CHP *International Journal of Applied Engineering Research* **11**(19) pp 9940-9945
[17] Lisin E, Kindra V, Dudolin A 2020 Development of Highly Economical Thermal Schemes of GTU-CHP in the Design of Energy-Saving Power Systems In *2020 International Multi-Conference on Industrial Engineering and Modern Technologies (FarEastCon)* p 20256571

A new technique for modelling phonon scattering processes at rough interfaces and free boundaries of solids

A A Barinov[1], B Liu[2], and V I Khvesyuk[3]

Bauman Moscow State Technical University, 2-ya Baumanskaya ul. 5, Moscow, 105005 Russia

[1] barinov@bmstu.ru, [2] liubinbmstu@gmail.com, [3] khvesyuk@bmstu.ru

Abstract. Scattering processes at interfaces and free boundaries of solids strongly affect heat transfer in micro- and nanostructures such as integrated circuits, periodic nanostructures, multilayer thin films, and other nanomaterials. Among many influencing factors, surface roughness due to atomic disorder plays a significant role in the rate of thermal transport. Existing approaches have been developed only for the limiting cases of smooth or completely diffuse surfaces. We have developed a new effective and simple method based on a direct consideration of the scattering of elastic waves from a statistically random profile (using a normal Gaussian surface as an example). This approach, first, allows to generalize common methods for determining the thermal properties of a real random rough surface using simple modifications, and, second, provides a tool for calculating the Kapitza conductance and the effective longitudinal thermal conductivity and studying the influence of roughness on heat transfer.

1. Introduction

The study of energy transfer processes in nanoscale electrical circuits with powerful specific heat generation is a wide range of complex problems, whose solution is extremely relevant and absolutely necessary, from both theoretical and practical points of view [1, 2]. Unlike macroscopic bodies, for which the classical laws of heat transfer are applicable, for micro and nanoscale structures, it is necessary to use completely different methods based on the statistics of quasiparticles called phonons—quanta of lattice vibrations. Over the past decades, there have been great advances in this area [3], but there are also a number of blind spots in fundamental the research that hinder the development of new thermal physics and hence nanoelectronics. In this paper we consider two modern problems of thermal transport of micro- and nanostructures:

1) heat transfer through the contact surfaces between two solids (through interfaces). A quantitative measure of the rate of heat transfer through interfaces is the Kapitza conductance (the reciprocal is the thermal boundary resistance (TBR).

2) Scattering of heat carriers (phonons) at the free boundary of a sample during the propagation of heat along the films (tubes). Quantitatively, this process is characterized by the effective longitudinal thermal conductivity (ETC).

In this case, the boundary of the sample is rough, i.e., close to the real one, and not smooth, as is usually assumed in classical thermal physics. At the moment, there are no reliable methods for calculating thermophysical properties (such as TBR and ETC) that would provide sufficient predictive modeling and control of the thermal regime of promising devices. To substantiate this thesis, we will focus on each of the above problems in more detail.

Content from this work may be used under the terms of the Creative Commons Attribution 3.0 licence. Any further distribution of this work must maintain attribution to the author(s) and the title of the work, journal citation and DOI.
Published under licence by IOP Publishing Ltd

1.1. Thermal boundary resistance (of interface)

The analysis of heat transfer through the contact surface of two solid bodies is a continuation of the classical problem of determining the Kapitza conductance (resistance) [4]. The task is extremely important, because, first, nanoelectronic circuits contain millions of contacts, and second, estimates show that thermal boundary contacts create the greatest total resistance to heat removal [5]. The existing theory of TBR calculation is based on two models: the acoustic mismatch model (AMM) [6] and the diffusion mismatch model (DMM) [7]. At the same time, it is important that in the AMM the interface is considered in the limiting case of a smooth surface, and in the DMM it is completely diffuse, whereas real interfaces have a complex structure, including roughness.

The effect of interface roughness on TBR has been investigated experimentally [8–11] and more recently using molecular dynamics [12–14] and *ab initio* [15] methods. Note, first, that these methods do not allow calculations for a previously known surface structure. Second, depending on roughness, phonons with different wavelengths will scatter differently: some diffusely, and some specularly. Therefore, further development of the computational model obviously requires the use of statistical methods to analyze the mechanism of phonon scattering on rough surfaces, and we have taken the first steps in the development of this method [16].

1.2. Effective thermal conductivity (films and tubes, in-plane)

Experimental studies of nanostructures (e.g., nanowires) have shown that roughness has a significant effect on heat transfer [17] due to the interaction of phonons with free surfaces. This is a decisive factor in the development of reliable nanotransistors, nanolasers, superlattices, and other devices.

The conditions in nanostructures are such that the mean free path (MFP) of phonons from one surface of films to another is significantly less than the average MFP of internal interaction processes in the sample. This is the so-called ballistic heat transfer regime [3]. The modern theory of thermal conductivity is based on the consideration of two limiting cases: completely diffuse scattering at the sample boundary and specular scattering, where the MFP is limited only by internal processes (phonon-phonon, on impurities, etc.) [18]. In this case, the transition from a purely diffuse regime to a specular one is carried out through the specular parameter p, which allows one to relate the MFPs of ballistic phonons for the two specified limiting cases [18].

The above method does not allow predicting the thermal properties of such structures, since the specular reflection parameter is used as a fitting parameter to ensure better agreement between theory and experiment [19, 20]. However, it is worth noting that in [21, 22], it is proposed to evaluate the parameter p as an integral (average) parameter for the sample. At the same time, it is not considered how the wavelengths of phonons, their polarization, and the temperature, geometry, and size of the sample will affect the nature the phonon scattering at a rough boundary (diffuse or specular reflection). Analysis shows that the question of the further development of the computational model remains open.

2. Mathematical model

Despite the fundamental differences between phonon scattering at rough interfaces (problem 1) and the interaction of phonons with free rough boundaries (problem 2), the method we propose allows us to develop a *unified computational technique*.

First, the method is based on the consideration of phonons as quasiparticles that obey Bose–Einstein (BE) statistics. Therefore, a number of properties of the carrier medium (crystal lattice) are required to describe phonons: dispersion relations and the density of states (DOS). In this paper, we use a well-proven approach to the application of polynomial approximation of experimental data [19]. The result is a nonlinear dispersion relation that characterizes the elastic properties of a real medium. We also note that the use of phonon statistics makes it possible to calculate the average values of the required thermal properties (thermal conductivity, Kapitza conductance) as a function of temperature.

First, we use the Landauer formalism, according to which the Kapitza conductance (thermal boundary resistance) is given by the expression

$$h_K = \frac{1}{3}\sum_j \int_0^{\omega_{\max,j}} \int_0^{\pi/2} C_{ph}(\omega,T) DOS_j(\omega) v_{g,j}(\omega) \alpha_{1-2,j}(\omega,\theta) \cos\theta \sin\theta d\theta d\omega \quad (1)$$

and the effective thermal conductivity is

$$\kappa = \frac{1}{3}\sum_j \int_0^{\omega_{\max,j}} \int_0^{\pi/2} C_{ph}(\omega,T) DOS_j(\omega) v_{g,j}(\omega) l_j(\omega,T,\theta) \sin\theta d\theta d\omega, \quad (2)$$

where ω is the phonon frequency, j is the phonon polarization, C_{ph} is the phonon capacity, $C_{ph} = \hbar\omega \cdot \partial f_{BE}/\partial T$; f_{BE} is the Bose–Einstein distribution function, $f_{BE} = \left[\exp(\hbar\omega/k_B T) - 1\right]^{-1}$; DOS_j is the density of state [19]; $v_{g,j}$ is the phonon group velocity; θ is the angle between the direction of incidence of a phonon and the normal to the surface; $\alpha_{1-2,j}$ is the coefficient of energy transfer from material 1 to material 2, and l_j is the mean free path of phonons. Note that expression (2) is given in a quasi-isotropic approximation [19]; in this case, the heat flux propagates in the film (tube) or in the so-called in-plane direction.

Thus, when calculating thermophysical properties, roughness is taken into account through the transfer coefficient $\alpha_{1-2,j}$ in expression (1) and through the mean free path l_j in equation (2). Next we will talk about the features of the calculation of $\alpha_{1-2,j}$.

Second, both problems are qualitatively different from each other. In problem 1, part of the energy from medium 1 is transferred through the interface to medium 2 (see Figure 1), and in problem 2, the energy does not leave the carrier medium. Figure 1 shows a schematic image of scattering on a rough interface between two bodies (solid 1 and solid 2) [16]. According to the theory of elasticity, a mode transformation occurs at the boundary [23]: an incident wave with index (0) is transformed into two reflected (1) and (2) and two refracted (3) and (4) waves. Letters denote polarizations: longitudinal (P) or transverse, where oscillations occur in the plane of propagation (SV) or across it (SH).

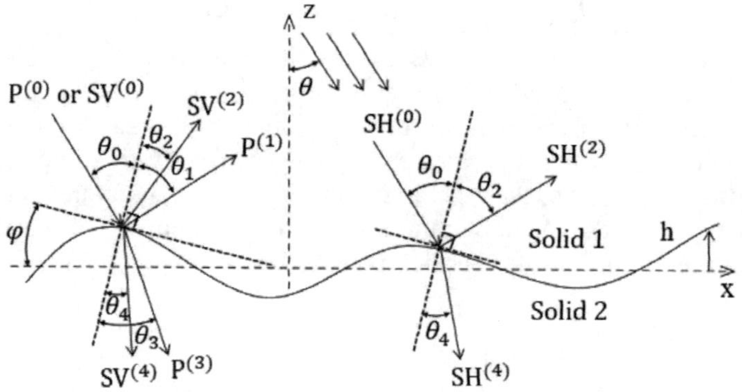

Figure 1. Scattering of elastic waves on a rough interface. The left picture corresponds to P or SV waves, and the right one to SH waves. Here the superscript (0) corresponds to the incident wave, (1) to the reflected longitudinal wave, (2) to the reflected transverse wave, (3) to refracted longitudinal wave, and (4) to the refracted transverse wave. The dashed line "x" is the mean line of the profile, and the solid line "h" shows the heights of the real (rough) profile.

Here we use the elastic wave model with dispersion (EWMD) to calculate the Kapitza conductance. The EWMD is a further development of the acoustic mismatch model (AMM), and it can be found in detail in [5, 16, 23].

Third, a distinctive feature of this method is that the description of the rough boundary of a sample is based on statistical accounting for the slopes of the real surface profile. For example, we consider a model of a two-dimensional random Gaussian normal surface $z = \zeta(x, y)$, which is characterized by two mean values: the root-mean-square roughness σ and the correlation length (interval) L. At the same time, unlike the well-known methods [18, 21, 22], the distribution density is used not for the profile heights z, but for profile slope gradients $n = |\nabla\zeta|$ [24]:

$$w(n) = \frac{1}{\gamma}\sqrt{\frac{2}{\pi}}\exp\left(-\frac{n^2}{2\gamma^2}\right), \quad (0 \le n < \infty), \tag{3}$$

where is the variance of the first derivative, $\overline{(\zeta')^2} \equiv \gamma^2$, which in the special case is given by the expression $\gamma = \sigma / l$. Note that in further calculations, as an argument we use not the gradient of the profile, but the angle of inclination φ (see Figure 1) between the mean line and the tangent plane, which are related as $n = \tan\varphi$.

Next, the problem of the wave propagation through the interface is solved relative to the tangent plane (see Figure 1). First, the transfer coefficient $\alpha_{1-2}(\theta_0)$ is determined as a function of the local incidence angle θ_0 using the well-known relations from the theory of elasticity [23]. Then using the relationship between θ_0, the angle of inclination of the profile φ at the scattering point, and the actual angle of incidence θ of the wave on the interface, we can pass from the local value $\alpha_{1-2}(\varphi,\theta)$ to the effective transfer coefficient $\alpha_{1-2,eff}(\theta)$ of the rough surface [16]:

$$\alpha_{1-2,eff}(\theta) = \int_{\varphi_{min}}^{\varphi_{max}} \alpha_{1-2}(\varphi,\theta) w(\varphi) d\varphi \tag{4}$$

It is due to the transition from α_{1-2} for the tangent plane to $\alpha_{1-2,eff}$ for the interface in expression (1) that the well-known relations of the theory of elasticity [23] can be naturally applied to the problem of reflection-refraction of elastic waves at a rough interface between two solids [16]. It should also be noted that when averaging over different angles of inclination of the profile (4), we need to take into account the restrictions on the limits of integration φ_{min} and φ_{max} [23] that are imposed by the phenomenon of total internal reflection and change the pattern of reflected-refracted waves.

The interaction of phonons with the rough wall of a sample during in-plane heat flux propagation is considered in a similar manner. The difference is that in expression (4), α_{1-2} is replaces with the MFP l_b between successive collisions with the boundary:

$$l_{b,eff}(\theta,h) = \int_{\varphi_{min}}^{\varphi_{max}} l_b(\varphi,\theta,h) w(\varphi) d\varphi \tag{5}$$

Details of the calculation of l_b can be found in the original work [18] or in our previous analysis [19]. After that, the average MFPs $l_{b,eff}$ are used to determine the effective longitudinal thermal conductivity of films (heat transfer rate) as a function of the size of the sample, its roughness, and temperature. The equation for ETC is similar to (2) but must include integration over k-space (reciprocal space of wavevectors) instead of integration over frequency. Moreover, to take into account the size effect in films, it is necessary to perform summation in the cross-plane direction and integration in the in-plane direction (for more information, see [25]).

3. Results and discussion

As an example, the problem of determining the Kapitza conductance through a rough aluminum-silicon interface is considered. The necessary properties and parameters of the materials are given in [16]. Here we focus on the calculation results presented in Figure 2. The initial conductance is the Kapitza conductance for a smooth interface (black line, "smooth"), which is obtained using the elastic wave model with dispersion (EWMD) [23]. Further, with an increase in roughness, a significant decrease in conductivity is observed. This is due to the fact that the proportion of waves capable of overcoming the interface and moving from medium 1 to medium 2 decreases, and, as a consequence, the effective transfer coefficient (4) decreases. For more information on the influence of wave polarization and interface roughness on the transfer coefficients see the article [16]. Note that here we are talking about inhomogeneities of an atomic-scale surface—the so-called atomic roughness.

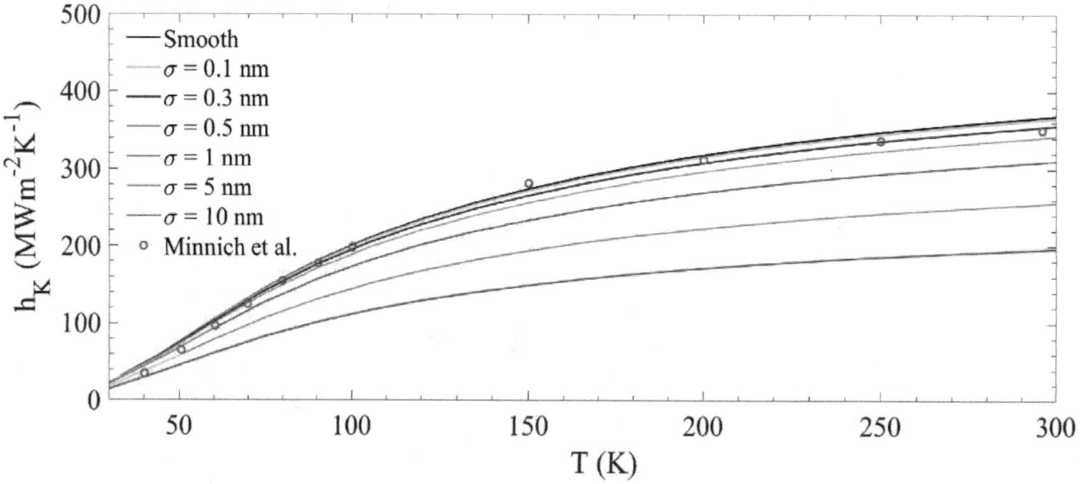

Figure 2. Kapitza conductance for various Al/Si interfaces: smooth and with a root-mean-square roughness of 0.1, 0.3, 0.5, 1, 5, and 10 nm. Circles represent experimental data [26].

Thus, we have developed a new technique for studying the effect of real surface roughness of a sample on the rate of heat transfer. Our method is the first to use the statistics of profile slopes of a random surface (using, as an example, the Gaussian distribution of slopes). The presented methodology, first, made it possible to formulate a statistical model for calculating the Kapitza conductance taking into account the roughness of the interface [16] and use it to calculate the silicon-aluminum contact (Figure 2). Second, it allowed us to develop a new method for calculating the effective thermal conductivity in the ballistic regime of heat transfer taking into account the multiparametric and statistical nature of phonons and a random (rough) surface.

References

[1] Moore A L and Shi L 2014 *Mater. Tooday* **17** (4) 163
[2] Pop E and Goodson K E 2006 *J. Electron. Packag.* **128** 102
[3] Khvesyuk V I and Skryabin A S 2017 *High Temp.* **55** (3) 428
[4] Kapitza P 1941 *J. Phys.(USSR)* 4 181
[5] Liu B and Khvesyuk V I 2020 *Int. J. Heat Mass Tran.* **159** 120117
[6] Khalatnikov I M 1952 *Zh. Eksp. Teor. Fiz.* **22** 687
[7] Swartz E T and Pohl R O 1989 *Rev. Mod. Phys.* **61** 605
[8] Blank M, Schneider G, Ordonez-Miranda J and Weber L 2019 *J. Appl. Phys.* **126** 165302
[9] Wang J, Zhang Z, Shi R, Chandrashekar B N, Shen N, Song H, Wang N, Chen J and Cheng C 2020 *Adv. Mater. Interfaces* **7** 1901582
[10] Zhao J, Zhao R, Huo Y and Cheng W 2019 *Int. J. Heat Mass Tran.* **140** 705

Journal of Physics: Conference Series **2150** (2022) 012021 doi:10.1088/1742-6596/2150/1/012021

[11] Ravichandran N K, Zhang H and Minnich A J 2018 *Phys. Rev. X* **8** 041004
[12] Termentzidis K, Parasuraman J, Da Cruz C A, Merabia S, Angelescu D, Marty F, Bourouina T, Kleber X, Chantrenne P and Basset P 2011 *Nanoscale Res. Lett.* **6** 288
[13] Termentzidis K, Merabia S, Chantrenne P and Keblinski P 2011 *Int. J. Heat Mass Tran.* **54** 2014
[14] Merabia S and Termentzidis K 2014 *Phys. Rev. B* **89** 054309
[15] Tian Z, Esfarjani K and Chen G 2012 *Phys. Rev. B* **86** 235304
[16] Liu B, Khvesyuk V I and Barinov A A 2021 *Phys. Solid State* **63** (7) 1128
[17] Lim J, Hippalgaonkar K, Andrews S C, Majumdar A and Yang P 2012 *Nano Lett.* **12** (5) 2475
[18] Ziman J M 1960 *Electrons and phonons. The theory of transport phenomena in solids* (Oxford University Press, New York) p 556
[19] Barinov A A, Liu B, Khvesyuk V I and Zhang K 2020 *Phys. Atom. Nucl.* **83** (10) 1538
[20] Turney J E, McGaughey A J H and Amon C H 2010 *J. Appl. Phys.* **107** 024317
[21] Soffer S B 1967 *J. Appl. Phys.* **38** 1710
[22] Maznev A A 2015 *Phys. Rev. B* **91** 134306
[23] Khvesyuk V I, Liu B and Barinov A A 2020 *Tech. Phys. Lett.* **46** 983
[24] Bass F G and Fuks I M 1979 *Wave Scattering from Statistically Rough Surfaces* (Oxford: Pergamon)
[25] Prasher R S and Phelan P E 1999 *Int. J. Heat Mass Tran.* **42** 1991
[26] Minnich A J, Johnson J A, Schmidt A J, Esfarjani K, Dresselhaus M S, Nelson K A and Chen G 2011 *Phys. Rev. Lett.* **107** 095901

Journal of Physics: Conference Series **2150** (2022) 012022 doi:10.1088/1742-6596/2150/1/012022

Problems of acoustic safety in power engineering

V B Tupov, A A Taratorin, V S Skvortsov and A B Mukhametov

Thermal Power Plants Department, National Research University "Moscow Power Engineering Institute", ul. Krasnokazarmennaya 14, Moscow, 111250 Russia

TupovVB@mail.ru

Abstract. Environmental safety issues are becoming increasingly important in the life of society. Among environmental safety issues in power engineering, acoustic safety occupies a special place. The problem of acoustic safety is associated with the fact that the regular operation of power equipment leads to an increased noise level, and power facilities are located in close proximity to residential areas. In this work, acoustic calculations were performed to determine the sanitary protection zone for gas turbines units (GTU) and combined cycle gas turbine units (CCGT) of various capacities. A formula was obtained for calculating the width of the sanitary protection zone depending on the capacity of gas turbine units and combined cycle plants and their number. It is shown that the sanitary protection zone (SPZ) of a power unit of high capacity is smaller than the sanitary protection zone of several power units of the same capacity. It is found that the noise levels from individual groups of equipment can determine the noise level at the entire border of the sanitary protection zone or in its individual sections. At the same time, noise suppression measures should be taken for all sources that generate noise levels in excess of standards. It is necessary to start noise suppression measures from those sources that generate excess noise in a larger section of the sanitary protection zone.

1. Introduction

Environmental safety issues are now becoming increasingly important in the life of society. Among environmental safety issues in power engineering, acoustic safety occupies a special place. The problem of acoustic environmental safety is associated with the fact that the regular operation of power equipment produces increased noise, and power facilities are located in close proximity to residential areas.

New requirements SanPiN 1.2.3685-21 "Hygienic standards and requirements for ensuring the safety and (or) harmlessness of environmental factors for humans" [1], introduced in March 2021, limit permissible noise values at the border of sanitary protection zones to 55 dBA during daytime and 45 dBA at nighttime. For power facilities, the boundaries of sanitary protection zones can be located at 300, 500, and 1000 m [2].

A feature of a power facility is a large number of sources of constant noise, their continuous operation, close proximity to residential areas, especially in large cities. Sources of continuous noise vary greatly depending on the type of equipment used. Strong sources of noise for the surrounding area are the air intakes of the blower fans and compressors of gas turbine units, the nozzles of gas ducts of smoke exhausters and pipes of gas turbine and combined cycle power plants, transformers, gas distribution points, cooling towers, coal-grinding equipment, the bodies of draft machines, and

Content from this work may be used under the terms of the Creative Commons Attribution 3.0 licence. Any further distribution of this work must maintain attribution to the author(s) and the title of the work, journal citation and DOI.
Published under licence by IOP Publishing Ltd

noise penetrating from noisy premises, etc. [3]. Axial draft machines are much more noisy than centrifugal ones. A classification of noise sources in turbine power plants (TPPs) is given in [3–4]. The most significant source of noise is the release of steam into the atmosphere. A large number of papers are devoted to the issues of noise from equipment and the creation of sanitary protection zones of power facilities [5–12]. Noise levels at the border of sanitary protection zones depend on the capacity and type of equipment used [11–12].A comparison of noise levels from steam power equipment with CCGT equipment is made in [12].

This paper presents calculations for both individual TPPs and individual groups of equipment. The width of sanitary protection zones of thermal power plants with advanced equipment for both gas turbine and CCGT units us discussed with regard to their capacity. In addition, the impact of equipment such as fan cooling towers and transformers is considered.

2. Results of calculations of SPZ aroundGTU and CCGT

The calculations were carried out for GTU and CCGT units of various capacities: GTU-78 MW and CCGT units with a capacity of 114 MW and 450 MW.

Acoustic calculations were performed in accordance with ISO 9613–2: 1996 [13]. In this case, the individual sound power levels of each piece of equipment, noise propagation conditions, and climatic and terrain features are taken into account. Such complex calculations are performed using specialized software products. The calculation is carried out according to the formula [13]:

$$L = L_W + D_C - A, \quad (1)$$

where L_W is the octave-band sound power level in decibels produced by the point sound source relative to a reference sound power of 1 picowatt (1 pW); D_C is the directivity correction in decibels that describes the extent to which the equivalent continuous sound pressure level from the point sound source deviates in a specified direction from the level of an omnidirectional point sound source producing sound power level L_W; A is the octave-band attenuation in decibels that occurs during propagation from the point sound source to the receiver.

The octave-band attenuation calculating by the formula

$$A = A_{div} + A_{atm} + A_{gr} + A_{bar} + A_{misc}, \quad (2)$$

where A_{div} is the attenuation due to the geometrical divergence (the energy divergence upon emission into free space); A_{atm} is the attenuation due to atmospheric absorption; A_{gr} is the attenuation due to the ground effect; A_{bar} is the attenuation due to a barrier; A_{misc} is the attenuation due to miscellaneous other effects.

The use of a specialized program for acoustic calculations makes it possible to simulate the location of TPP buildings with their overall dimensions with the main sources of noise. In the calculations, the shielding effect from buildings and structures of TPPs is taken into account. Acoustic calculations using a specialized program allow one to determine the noise level both from the main sources of noise and from individual sources or groups of sources at a distance from the TPP taking into account the relative position of sources in the area of the station and their height above the ground. The latter is especially important since artificial and natural barriers do not have a shielding effect on the radiation from a high-altitude source.

To determine the noise levels at the border of the sanitary protection zones, mathematical modeling of TPP with gas turbine equipment and CCGT was performed. At the same time, dry fan cooling towers are used at some stations, and natural draft counterflow cooling towers are installed at other TPPs.

Figure 1 shows the results of calculations of the sanitary protection zone for operation of a 78 MW GTU and a 114 MW CCGT and a 450 MW CCGT units. The number of operating units was varied from 1 to 4. Acoustic calculations were performed for the entire set of equipment operating with the specified gas turbine and CCGT units in the factory configuration without additional measures for noise suppression. In all cases, natural draft counterflow cooling towers are used. The operation of the

GTU and CCGT units requires the establishment of sanitary protection zones significantly exceeding the regulatory limits.

The calculation results in Figure 1 can be approximated by the formula for calculating the size of the sanitary protection zone *R*, m, from the electric power of the station *N*, MW:

$$R(N)=A\ \ln(N)-C, \quad (1)$$

where *N* is the electric power of the station, MW; *A* and *C* are empirical coefficients, which are presented in Table 1. The confidence level R^2 for the obtained values according to formula (1) is about 0.99.

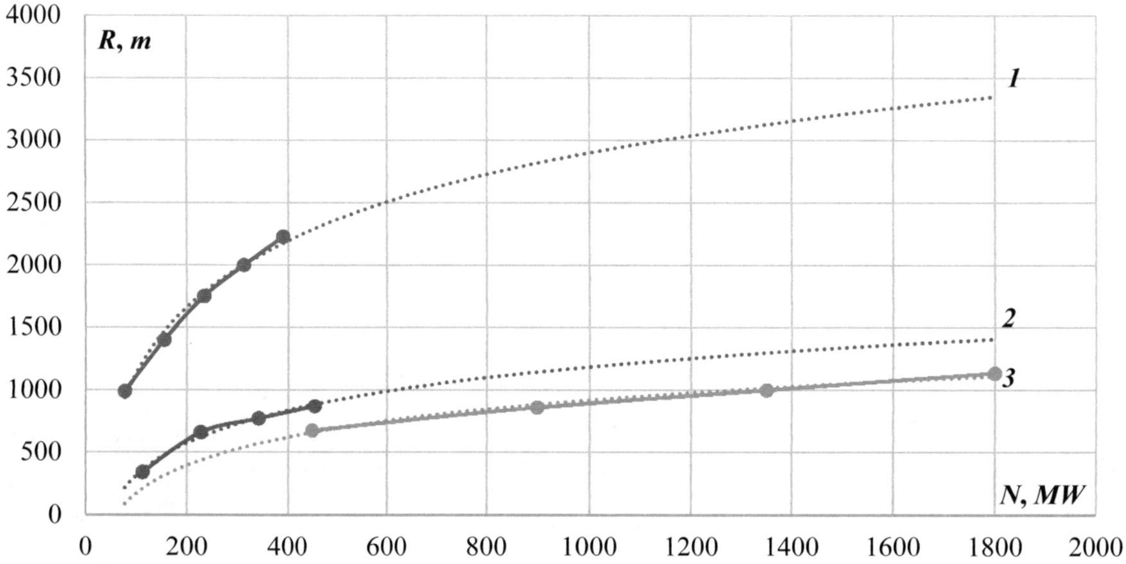

Figure 1. Width of the sanitary protection zone in a residential area versus the power of the station and the type of units to comply with the limit of 45 dBA for the boundaries of sanitary protection zones: 1 – GTU 78; 2 –CCGT 114; 3 –CCGT 450

Figure 1 shows that with the commissioning of a second unit, the sanitary protection zone increases by 535 m for a 78 MW GTU , by 263 m for a 114 MW CCGT unit, , and by 226 m for a 450 MW CCGT. For a TPP with a capacity of approximately 450 MW, the sanitary protection zone with the installation of one 450 MW CCGT unit is 674 m, with the installation of four 114 MW CCGT units, it is 870 m, and with the installation of five 78 MW GTU units,2314 m. The plant with five 78 MW GTU units, should have a sanitary protection zone 3.4 times larger than the TPP with one 450 MW CCGT unit. Thus, when choosing the composition of the equipment, one should give preference to equipment with a larger unit capacity.

The acoustic calculations performed make it possible to determine the main sources of noise responsible for the exceedance of the sanitary limits in the area surrounding the sanitary protection zone. For example, for supercritical pressure units, this is the noise from exit nozzles of axial smoke exhausters [3–4]; for TPPs with gas turbine units, this in some cases may be the noise from dry fan cooling towers [14].

Table 1. Empirical coefficients characterizing the dependence of the size of the SPZ on the electrical power for different types of units

Units	Coefficient A	Coefficient C	Confidence R^2
GTU 78	767,79	2406.2	0.99
CCGT 114	379.97	1440.9	0.99
CCGT 450	326.22	1334.6	0.99

3. The results of calculations for individual groups of power equipment

Tthe results of calculations for the main sources of noise for various sections of the sanitary protection zone are given below. For example, the section of the sanitary protection zone may be large for dry fan cooling towers and small for transformers.

Consider the noise from fan cooling towers. It has been shown [14] that fan cooling towers are always noisier than natural draft counterflow cooling towers for cooling the same amount of water from condensers. Therefore, the use of high-capacity dry fan cooling towers at TPPs instead of natural draft counterflow cooling towers increases the noise level in the surrounding area.

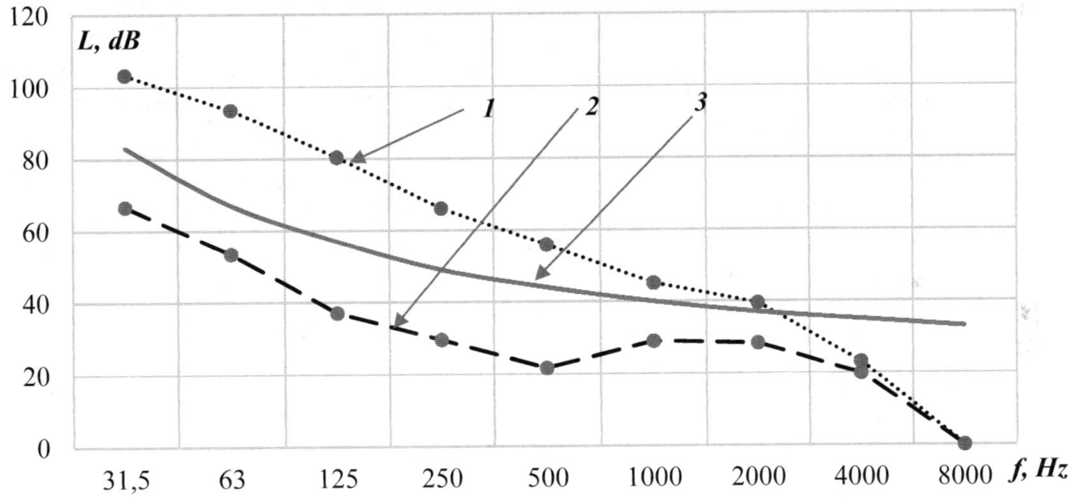

Figure 2. Sound pressure levels at the design point on the border of the sanitary protection zone during operation of natural draft cooling towers and fan cooling towers of a 78 MW GTU: 1–dry fan cooling towers; 2 –natural draft counterflow cooling towers; 3 –sanitary standards at 45 dBA

Figure 2 shows the maximum sound pressure levels at the design point at the border of the sanitary protection zone during operation of natural draft counterflow cooling and dry fan cooling towers required for the operation of a 78 MW GTU. The sound pressure levels during operation of dry fan cooling towers will be higher than sound pressure levels during operation of natural draft counterflow cooling towers for low- and medium-octave band frequencies. For example, the difference in sound pressure levels between a natural draft counterflow cooling tower and a dry fan cooling tower will be about 40 dB for octave band frequencies of 31.5, 63, and 125 Hz. During the operation of fan cooling towers, the sanitary limits will be exceeded at the control point at the boundary of the sanitary protection zone for octave band frequencies from 31.5 to 1000 Hz.

Figure 3 shows the maximum sound pressure levels at the design point on the border of the sanitary protection zone during the operation of transformers at thermal power plants with a capacity of

900 MW. For octave band frequencies of 1000 and 2000 Hz, the noise from the transformers is the main one. For other points shielded by buildings, the sound pressure level from the transformers does not exceed the norm.

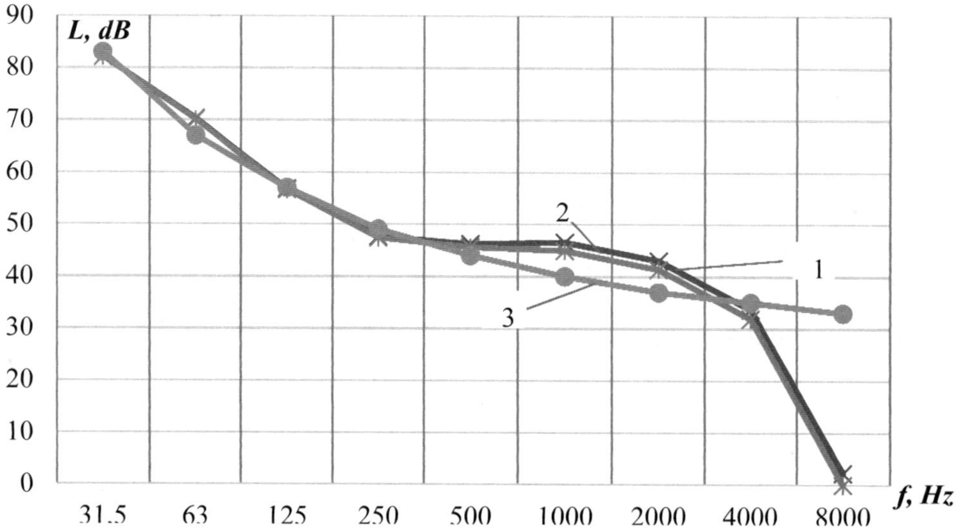

Figure 3. Sound pressure levels at the design point on the border of the sanitary protection zone during TPP operation: 1 – from all sources; 2 – from transformers; 3 – sanitary standards at 45 dBA

Individual groups of sources can affect noise levels in the entire surrounding area. These sources include steam emissions from the nozzles of gas ducts of smoke exhausters and pipes of GTU and CCGT. Most of the boundary of the sanitary protection zone is influenced by sources such as fan cooling towers and air intakes of blower fans. Some parts of the sanitary protection zone are affected by noise from transformers and hydraulic fracturing.

In any case, if the sanitary standards are exceeded, it is necessary to take measures to suppress noise. Possible measures to reduce the power of equipment are given in [15–17].

When developing measure to reduce the power and suppress the noise of power equipment, it is necessary to consider the regional climate factor [18– 20].

4. Conclusion

- The noise from energy facilities depends on both the type and power of the energy units of which it is composed. Increasing the capacity of an energy facility leads to an increase in the sanitary protection zone. At the same time, the noise level at the SPZ border is often determined by a separate group of sources.
- Formula (1) was obtained for calculating the width of a sanitary protection zone depending on the capacity of GTU and CCGT units and their number. It is shown that the sanitary protection zone of a power unit of high capacity is smaller than the sanitary protection zone of several units of the same capacity.
- It is shown that the noise levels from individual groups of equipment can determine the noise at the entire border of the sanitary protection zone or at its individual sections. At the same time, noise suppression measures should be carried out for all sources that generate noise in excess of standards. It is necessary to start noise suppression measures from those sources that generate excess noise in a larger section of the sanitary protection zone.

References

[1] SanPiN 1.2.3685-21 2021 *Hygienic standards and requirements for ensuring the safety and (or)*

harmlessness to humans of environmental factors. Moscow, [In Russian]

[2] SanPiN 2.2.1/2.1.1.1200–03 2003 *Sanitary protective zones and sanitary classification of enterprises, constructions and other objects* [In Russian]

[3] Tupov V B 2012 *The Factors of Physical Impact of Thermal Power Plants on the Environment* (Moscow: MPEI Publ.) [In Russian]

[4] Tupov V B 2017 *Increasing safety of thermal and nuclear power stations energy equipment by reducing noise* IOP Conf. Series: Journal of Physics: Conf. Series 891

[5] Panteli, M., Mancarella, P. *The grid: Stronger, bigger, smarter?: Presenting a conceptual framework of power system resilience* 2015 IEEE Power and Energy Magazine, 13 (3), art. no. 7091066, pp. 58-66

[6] Naik K, Mahapatra A 2013 *Assessment of noise pollution of Vedanta thermal power plant* Jharsguda, Odisha, India. International Journal of Chemical Sciences 11, 2: 1016-1028.

[7] Komarova E G, Pimenov I K 2014 *Specificity of the application of the inverse method for calculating the acoustic evaluation of sanitary protection zones of industrial enterprises.* Technical Acoustics **5**

[8] Chernosvitov M D, Gorshkalev P A, Porshina E G 2019 *Sanitary Protection Zones of Sewage Waste-Disposal Plants and Their Justification.* In IOP Conference Series: Earth and Environmental Science, vol. 272, **2**, p. 022253. IOP Publishing

[9] Fan X, Li L, Zhao L, He H, Zhang D, Ren Z, Zhang Y. 2020 *Environmental noise pollution control of substation by passive vibration and acoustic reduction strategies* Applied Acoustics, 165, August 2020, 107305

[10] Tupov V B, Taratorin A A 2020 *Features of noise radiation from gas turbines.* Journal of Physics: Conference Series (JPCS).1683 042089

[11] Tupov V B, Taratorin A A 2020 *Influence of thermal power equipment structure on the size of sanitary protection zone by noise factor* Proceedings of the 2st IEEE 2020 International Youth Conference on Radio Electronics, Electrical and Power Engineering, REEPE 2020. Paper 9059202

[12] Tupov V B, Taratorin A A 2020 *Influence of replacement of steam power plant on combined cycle gas equipment on the sanitary protection zone by noise factor* Journal of Physics: Conference Series, 1683(4), 042070

[13] ISO 14044: 2006, *Environmental Management—Life Cycle Assessment—Requirements and Guidelines*. International Organization for Standardization: Geneva, Switzerland

[14] Tupov V , Taratorin A , Mukhametov A. 2021 *Comparing the noise characteristics of different types of cooling towers* Akustika, **39**, 44–47.

[15] Tupov V B 2013 *A package of measures to reduce the noise from thermal power plants* Power Technology and Engineering 47, **3**, 217-221

[16] Tupov V and Tupov B 2017 *Solutions to creating a low-noise power facility* 24th International Congress on Sound and Vibration, London, UK

[17] Semin S, Tupov V 2010 *Optimal design of dissipative silencer for gas turbine noise reduction* 39th International Congress on Noise Control Engineering **8**, 6043-6052

[18] Tupov V B, Taratorin A A, Skvortsov V S 2018 *Effect of Regional Climatic Factors of Reducing Noise Level from Power Equipment.* Thermal Engineering 65, **11,** 841-845

[19] Tupov V and Taratorin A 2013 *Influence of climatic factors and the ground surface on the required noise abatement from power equipment* Thermal Engineering 60,**7,** 493-497

[20] Tupov V, Taratorin A., Kuzminova S, Skvortsov V. 2021 *Results of Measures Aimed at Reducing Noise from a Gas Distribution Plant and Adjacent Gas Pipelines.* Power Technology and Engineering, 55, **2**, 279-281

Properties of anhydrous hydrogen fluoride in the temperature range from 140 K to 190 K

S M Gubanov

National Research Tomsk State University, Tomsk, 634050 Russia

Sgubanov@yandex.ru

Abstract. This paper presents the analysis of the known empirical formulas on the pressure of the saturated vapors of hydrogen fluoride depending on the temperature. The possibility of using the formulas to determine the vapor pressure under the low temperatures is assessed. The calculation error of the pressure value depending on temperature outside the interval are determined.

1. Introduction

The separation technology of uranium isotopes includes the purification of the final product from light impurities, which is mostly hydrogen fluoride (HF). The purification is usually carried out by the desublimation of HF from the gas mixture stream into cooled containers using the special equipment. The main purpose of the desublimation plants is to eliminate the emission of the harmful substances into the atmosphere. Hydrogen fluoride formed during the production process is recovered from the gas mixture. The rest of the components of the mixtures are filtered and neutralized according to the sanitary standards before being discharged into the atmosphere. The properties of the anhydrous HF are: melting point - 190 K, boiling point - 292.5 K, and density - 0.99 g/cm^3. It dissolves well in water but is flammable and explosive. The average daily maximum permissible concentration in the air of the settlements is 0.005 mg/m^3. The critical concentration value is 8 mg/m^3.

The condensation-evaporation section is designed to condense the UF_6 residues and to purify the process stream from the light and harmful impurities. Fig. 1 shows the schematic diagram of the gas mixture separation process. The temperature level of 253 K is used to condense the bulk of the UF_6. The desublimation unit is thermostated at T = 193 K by air, and at T = 77 K - by liquid nitrogen (Fig. 1).

At the last stage, HF desublimation takes place in special containers cooled by liquid nitrogen. This cooling method is expensive and labor intensive. The main purpose of the desublimation plants is to eliminate the emission of the harmful substances into the atmosphere.

The developed chiller BXM-0,54/0,6 [1] allows cooling containers at the last stage of the desublimation by air in the temperature range of 130 K÷140 K [2]. To estimate the amount of uncondensed HF, it is necessary to measure the pressure of the saturated vapor in this temperature range. The empirical formulas known in the scientific literature [3-5] were obtained by direct measurements at a higher temperature, which is above the melting point of hydrogen fluoride. Such measurements for the temperature range from 140 K to 190 K were performed in [5].

The aim of this study is to determine the residual amount of HF in the process stream after the desublimation at the last stage when the tanks are cooled by cold air at the temperature of 130 K ÷ 140 K.

Content from this work may be used under the terms of the Creative Commons Attribution 3.0 licence. Any further distribution of this work must maintain attribution to the author(s) and the title of the work, journal citation and DOI.
Published under licence by IOP Publishing Ltd

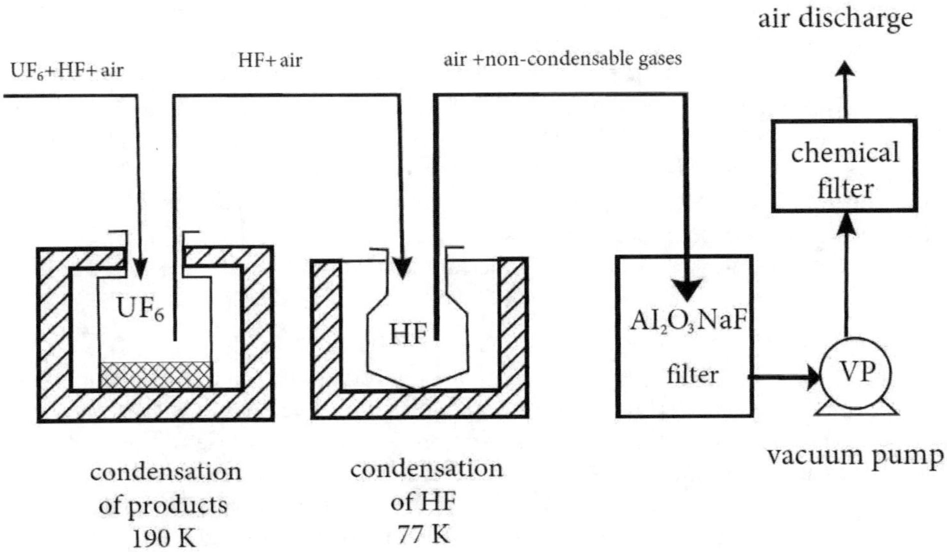

Figure 1. Fractional separation of the gas mixtures

2. Formulation of the problem

After the development and establishment of the BXM-0,54/0,6 serial production [1], the HF desublimation section was switched to air cooling, which is less energy-intensive [2]. To determine the amount of HF passing through the section of its desublimation, it is necessary to determine the pressure of its saturated vapor at the temperature of the cold air in BXM-0,54/0,6.

The formulas given in [3-5] has not given a satisfactory result of the vapor pressure. The temperature ranges in which formulas [3-5] are reliable are higher than those realized in the air chiller. Formula [3] has been obtained in the temperature range from 192 K to 320 K, formula [4] - in the range from 273 K to 378 K, and formula [5] - from 293 K to 473 K. The pressure values in the temperature range from 140 K up to 190 K is not correct due to the extrapolation of the dependences to the temperature values in which measurements have not been carried out.

3. Solution

One of the first data on the saturated vapor pressure of HF at low temperatures was published in 1924 by J. Simons [3]. In the experiments, hydrogen fluoride was frozen by the liquid air (at 90 K) with subsequent heating. The measurement data were processed as the dependence of the vapor pressure on the melting point in the temperature range from 192 K to 320 K:

$$\log(P) = 7.37 - 1315 / T, \tag{1}$$

where P is the pressure, mm Hg.

In 1953, an article was published [4]. The vapor pressure of HF was measured in the temperature range from 273 K to 378 K:

$$\log(P) = 8.38036 - 1952.55 / (62.37 + T) \tag{2}$$

where P is the pressure, mm Hg.

Paper [5] provides data on the saturated vapor pressure of HF at high pressures in the temperature range from 293 K to 473 K. The obtained formula is:

$$\ln(P / 1000) = 25.5266 - 3267.53 / T - 1.82355 \cdot ln(T) + 7.07191 \cdot 10^{-6} T^2 \tag{3}$$

where P is the pressure, kPa.

Formulas (1) - (3) determine the saturated vapor pressure of HF above the surface of the liquid anhydrous HF. The authors in [7] conducted the analysis on the influence of the HF molecules association on the value of the saturated vapor pressure. It was noted that the degree of the molecule association increases with decreasing temperature.

The formulas (1) - (3) are limited by three factors, which do not allow determining the saturated vapor pressure of HF at the temperatures below 190 K (HF solidification temperature): 1) in the formulas (1) - (3), the association of HF molecules with decreasing temperature is not taken into account; 2) formulas (1) - (3) were derived for the saturated vapor pressure above the liquid surface; 3) the extrapolation operation of the formulas to the temperature values outside the measurement interval is not a correct mathematical operation.

The theoretical temperature dependence of the saturated vapor pressure has the form:

$$P = A\exp\left(-\frac{L}{RT}\right), \tag{4}$$

where L is the phase-transition heat, R is the universal gas constant, and A is the constant.

It is known that during the sublimation and desublimation, the heat of the phase transition differs from the one during the evaporation and condensation. Therefore, the use of the formulas (1) - (3) to determine the pressure of saturated vapor over a solid HF surface is fundamentally impossible.

To determine the pressure of the saturated HF vapor over its solid surface, we have carried out the measurements on the specially designed stand [6]. The obtained formula to determine the dependence of the HF saturated vapor pressure on the pressure in the temperature range of 140K ÷ 190K is:

$$P = 1.4879\cdot 10^{11}\exp\left(-3911.2/\mathrm{T}\right), \tag{5}$$

where P is the pressure, Pa.

Fig. 2 shows the calculation result of the HF saturated vapor pressure by the formulas from [3-6] in the temperature range of their derivation. It is noteworthy that the slope of the lines 1, 2, 3 is practically the same, but the slope of the line 4 is different. This indicates that the value of the phase transition heat in the formula (5) differs from the value determined by the formulas (1) - (3).

Fig. 3 shows the calculation results on the HF saturated vapor pressure according to the formulas from [3-6] in the temperature range from 140 K to 160 K. According to the figure, the saturated vapor pressure calculated by the formulas [3-5] significantly overestimates the values obtained from the experiments and approximated by the formula (5) [6].

The authors in [8] present the mathematical model and the calculation results on the HF desublimation on the inner walls of the tank. The calculation results according to the model [8] are plotted in Figs. 4, 5. Fig. 4 shows the distribution of the HF concentrations in the air along the height of the tank when it is cooled by the BXM at the temperature of 113 K, with the 6th level of the molecule association. During the residence time of the mixture in the tank, it cools down to the chiller temperature, having passed about the half of the tank height (Fig. 5). At the same distance, the saturated vapor pressure HF takes on the pressure value corresponding to the wall temperature.

We have estimated the HF mass concentration in the flow outgoing from the tank. The absolute pressure of the mixture is 50 Pa. The partial HF pressure is 45 Pa, and air - 5 Pa. When the mixture flow cools down to the temperature of 140 K, the partial HF pressure according to the formula (5) is 0.1067 Pa. The ratio of the partial pressures of HF and air and the ratio of the number of HF molecules to the number of air molecules are:

$$\frac{P_{HF}}{P_b} = \frac{n_{HF}}{n_b} = 0.00213,$$

where P is the partial pressure, and n is the molecule number per unit volume. Indexes are *HF* is hydrogen fluoride and b - air.

The mass concentration of HF in the gas mixture has the value:

$$\frac{\rho_{HF}}{\rho_b} = \frac{n_{HF}\mu_{HF}m}{n_b\mu_b} = 0.00213\frac{19\cdot 6}{29} = 0.00837\frac{\text{кг}(HF)}{\text{кг}(b)},$$

where μ is the molar mass, ρ is the density, and m is the molecule association level.

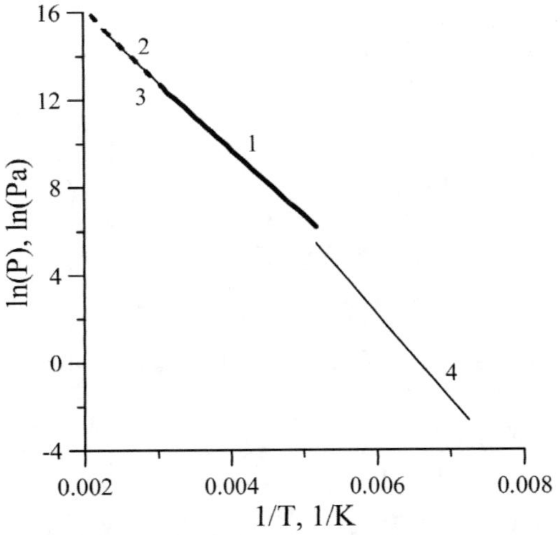

Figure 2. HF saturated vapor pressure calculated by the formulas: 1 – (1) from [3], 2 – (3) from [5], 3 – (2) from [4], 4 – (5) from [6].

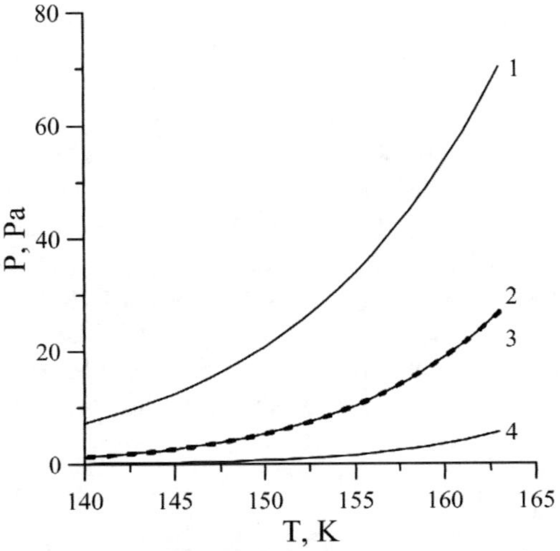

Figure 3. HF saturated vapor pressure in the temperature range of 140 K ÷ 160 K, calculated by the formulas: 1 – (1) from [3], 2 – (3) from [5], 3 – (2) from [4], 4 – (5) from [6].

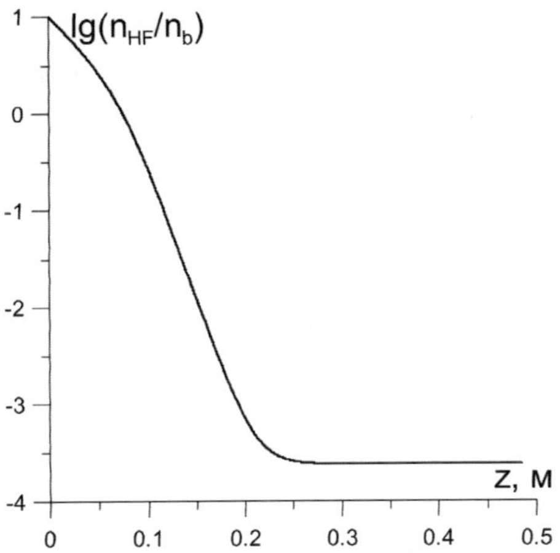

Figure 4. HF concentrations in the air along the height. $T_w = 113\ K$, molecule association level m=6.

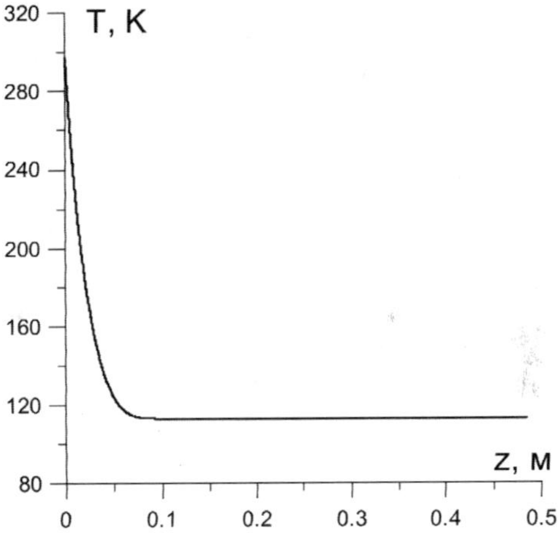

Figure 5. Temperature distribution along the height. $T_w = 113\ K$, molecule association level m=6.

Based on the obtained data, it is possible to estimate the required capacity of the chemical HF absorber that meets the technology requirements.

4. Conclusion

We have presented the analysis of the known empirical formulas on the temperature dependence of the pressure of the HF saturated vapors and its application at the low temperatures outside the empirical temperature range. The paper provides data on the calculation error of the HF saturated vapor pressure for the case of temperatures going beyond the range of the formulas' derivation. We have also determined the HF mass concentration in the outgoing flow of the gas mixture from the desublimation tank by the empirical formula for the pressure of saturated vapor over the HF solid surface.

References:

[1] Gubanov S M, Udut V N, Stulov V L 2016 *Patent* 97784 Russia
[2] Gubanov S M, Krainov A Y, Mazur R L, Durnovtsev M I 2017 *Theoretical Foundations of Chemical Engineering* **51** No 3 352-358
[3] Simons J 1924 *J Am Chem Soc* **46** No 10 2179-2183
[4] Jarry R L and Davis W Jr 1953 *J Phys Chem* **57** No 6 600– 604
[5] Wilson L C, Wilding W V and Wilson G M 2014 *J. Chem Eng Data* **59** No 4 983–990
[6] Gubanov S M, Durnovtsev M I, Krainov A Y 2017 *Journal of Engineering Physics and Thermophysics* **90** No 3 634-636
[7] Vetere A 1991 *Ind Eng Chem Res* **30** 2487-2492
[8] Vasenin I M, Gubanov S M, Durnovtsev M I, Krainov A Y, Chukanov M V 2014 Tomsk State University Journal of mathematics and mechanics **31** No 5 76-82.

Power units with full CO2 utilization

O O Milman[1,2]

[1] CJSC SPC «Turbocon», Russia, Kaluga, Komsomolskaya roscha st. 43, 248010
[2] Kaluga State University named after K.E. Tsiolkovski
26, Stepan Razin str. Kaluga, 248023

turbocon@kaluga.ru

Abstract. The forecasts for the development of renewable energy and conventional energy with fossil fuels have significant discrepancies in quantitative indicators, but agree on the need to reduce CO2 emissions. In this direction, a great number of developments are associated with hydrogen energy. Alternative proposals are cycles on methane-oxygen fuel with CO2 capture from the concentrated stream at the outlet of the condenser-separator: high-temperature gas-steam turbine unit of CJSC SPC «Turbocon», Allam, and JIHT RAS cycle. A 100 kW gas-steam turbine with an initial mixture temperature of up to 800° C has been developed and tested. To develop a method for calculating steam condensers from a mixture with CO2 (common to all three schemes), tests were carried out on a special stand; the heat transfer coefficient experimental data have been used to design a highly efficient steam condenser from a mixture with a converging flow path to maintain its high speed, a heat transfer coefficient of 2700 W/m2K was achieved. It is planned to create a prototype installation containing a steam boiler, a gas-steam turbine and a steam condenser with CO2 capture from the concentrated stream at the compressor outlet.

Currently, the prospects for the development of energy up to 2040-2050 are actively discussed against the background of environmental problems and forecast of climate change. In the forecasts of energy consumption and electricity production, there is a significant variation, which can be associated with certain aspects. So, Figure 1 shows the energy consumption charts (a) and the production of electricity from renewable energy sources (RES) (b) [1]. The growth of resource consumption and moderate growth of RES are predicted by BP and Shell oil refineries. International Energy Association (Paris) and Irena Get (Abu Dhabi), where renewable energy is intensified, predict reducing energy consumption and rapid growth of renewable energy. Nevertheless, both sides of the discussion, differing in quantitative estimates of the transition process, recognize its inevitability and, as a result, the presence of signs of the IV energy transition and the increase in the share of renewable sources in general energy balance.

Thermal power engineering and transport are the most important anthropogenic sources of mass emission of carbon dioxide. The struggle for the reduction of these emissions is carried out in three directions:

- transition to "pure" types of fuel (hydrogen, natural gas);
- increasing the efficiency of thermodynamic processes (more energy, less emissions);
- calming or disposal of carbon dioxide at the exit of power unit.

Content from this work may be used under the terms of the Creative Commons Attribution 3.0 licence. Any further distribution of this work must maintain attribution to the author(s) and the title of the work, journal citation and DOI.
Published under licence by IOP Publishing Ltd

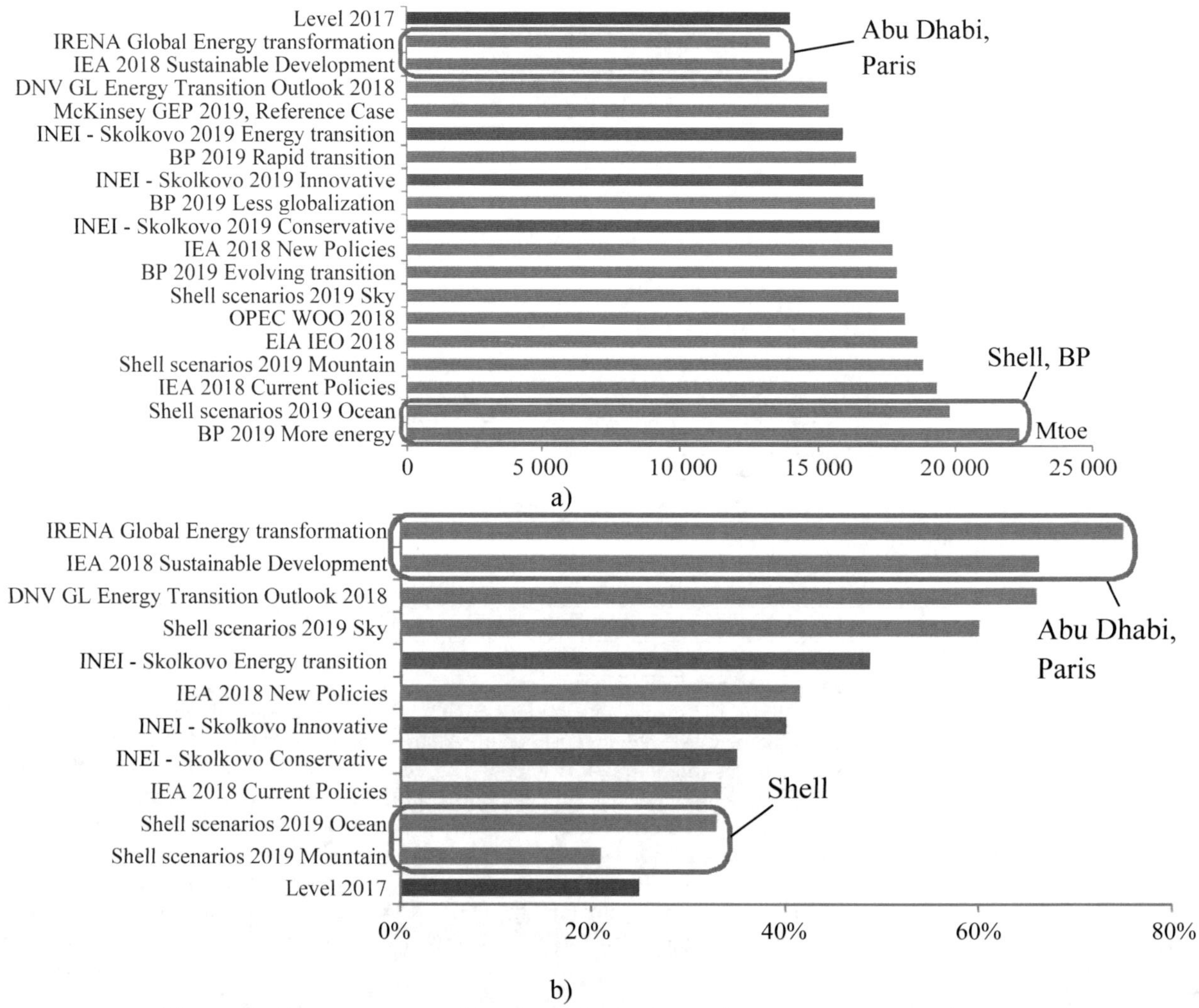

Figure 1. Energy consumption charts for traditional energy resources (a), electricity generation in RES (b) (2040).

In modern scientific and technical prospects marked by the Russian Government Resolutions, the focus is on the first direction. Similar trends take place abroad. Two other areas are equally important: an increase in the efficiency of energy complexes and economically justified systems of capture and disposal of CO_2.

The technological basis of the gas transition is illustrated in Figure 2. Among those proposed in Figure 2, hydrogen energy, being the cleanest in terms of hydrogen production, transportation and storage, remains very vulnerable.

In addition to the absence of CO_2 emissions, hydrogen fuels can significantly increase the initial temperature of water vapor in steam-turbine complexes and, as a result, to increase significantly their efficiency.

The fundamental technical solutions based on such units were formulated by E.E. Spielrain [2] in the 80s of the 20th century. The main idea is the combustion of hydrogen and oxygen in the water vapor environment after the classic steam boiler.

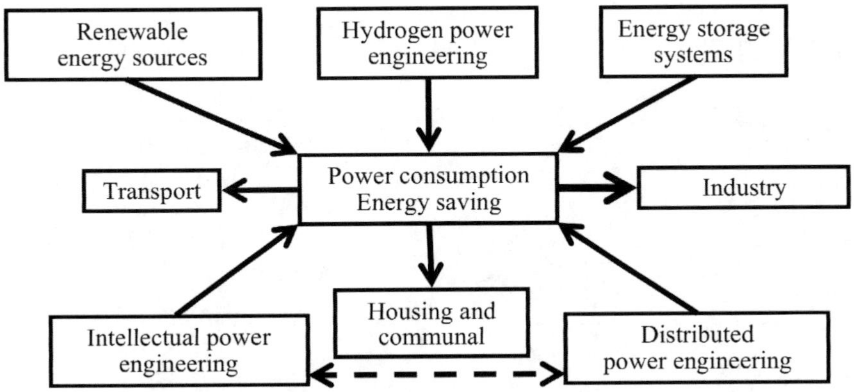

Figure 2. Technological basis of energy transition.

A cycle with a hydrogen overheating of steam is shown in Figure 3a, and the T-S diagram is shown in Figure 3b. The latter illustrates an important feature of hydrogen overheating: the cycle consists of two components: a classic "boiler version" with a flow rate G = 1, in which the steam turbine regeneration system plays significant role in increasing the efficiency, and the cycle of additional fuel G = α, where there is no regeneration, and the effectiveness of which is significantly lower than that of the "boiler room".

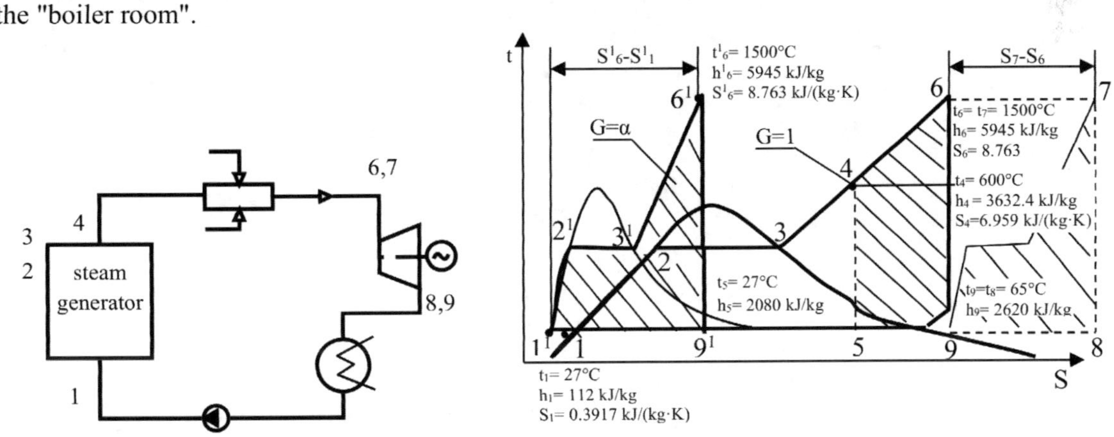

a) b)

Figure 3. Hydrogen superstructure diagram (a) and its thermodynamic cycle (b).

The development of this idea was the burning of natural gas and oxygen in the steam moving from the boiler (Figure 4). There are no problems specific to hydrogen overheating, but a mixture of steam and dioxide comes into the condenser. After condensation, we obtain substantially pure carbon dioxide, which can be recycled. In this case we can have CO_2 utilizing only for additional fuel.

As a result of the work in this direction, an experimental sample of a high-temperature installation of a power of 100 kW for steam parameters p = 3.4 MPa and T_0 = 800°C was created. Figure 5 shows the cut of the high-temperature steam turbine unit HSTU-100.

The two-rim Curtis stage and the three pressure stages provide operation of the unit at 24,000 rpm and pressure in the condenser of 10 kPa. The project of a high-temperature unit with a capacity of 300 MW for parameters P_0 = 30 MPa, T_0 = 1250°C/1450°C, with the net efficiency (minus a booster and air separation unit) amounted to 54% was developed together with the NRU «MPEI».

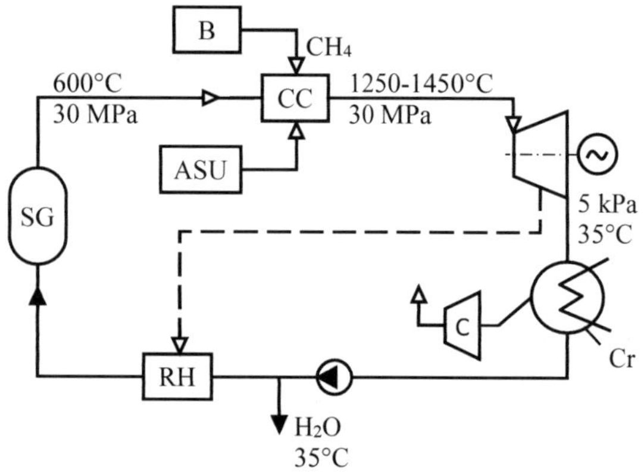

Figure 4. Gas and steam turbine plant with a gas steam superheating: SG – steam generator; B –booster, RH – regeneration heater; ASU – air separation unit; CC – gas overheating combustion chamber, Cr – condenser, C – CO_2 compressor.

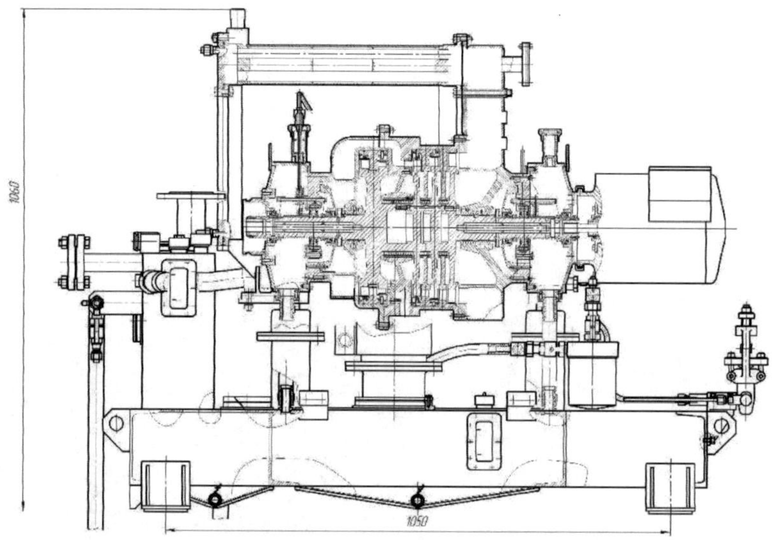

Figure 5. High-temperature steam turbine HST-100

At the heat electro power station HES-16 PJSC "Mosenergo" a system of gas steam overheating using methane - oxygen fuel was tested, a steam temperature of 720°C was achieved, and the performance of gas overheating technology was confirmed. The photo of the gas overheating unit before installing on the HES-16 is shown in Figure 6.

Other works in this regard are represented by the Allam cycle (Figure 2), the JIHT RAS cycle idea is similar. These two series share a common working body: carbon dioxide.

A schematic generalized cycle of Allam and JIHT RAS is shown in Figure 7. At a mixture temperature of 1100°C and a pressure in the circuit of up to 30 MPa, the authors of the cycle predict an efficiency of up to 54%.

When analyzing the Allan cycle, an estimate was made of the dimensions of the heat regenerator operating at temperatures of up to 800°C and pressures of up to 30 MPa. The design of this apparatus is not presented in the literature, our estimate shows that the length of the heat exchange tubes will be 90 - 130 m with a counter-current flow of the spent working fluid (the mixture of CO_2 and H_2O, p = 3.0 MPa) and carbon dioxide into the combustion chamber (p = 30.0 MPa).

Transfer of hydrogen problems to the problem of separating carbon dioxide from water vapor in a

Journal of Physics: Conference Series **2150** (2022) 012024 doi:10.1088/1742-6596/2150/1/012024

Figure 6. Gas steam superheating before installation at HES-16 of PJSC "Mosenergo".

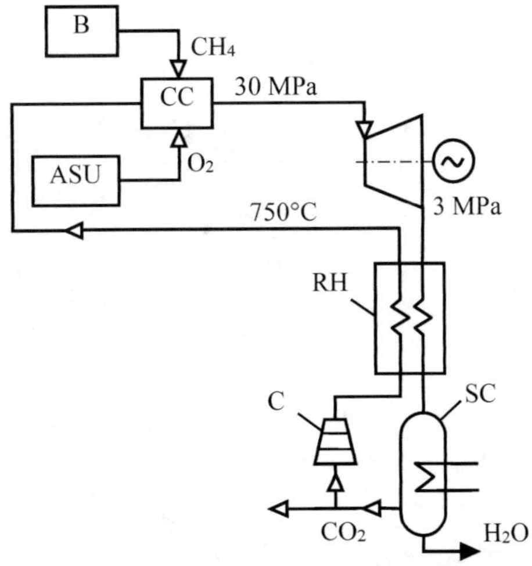

Figure 7. Power plant working on the Allam cycle [6]: SC - separator-condenser, other designations are according to Figure 4.

condenser/separator is common to all three cycles with natural gas and oxygen. To develop a methodology for calculating such devices, a tube bundle was tested on a special stand during condensation of water vapor from a steam-gas mixture (SGM). The design of the working section of the stand is shown in Figure 8.

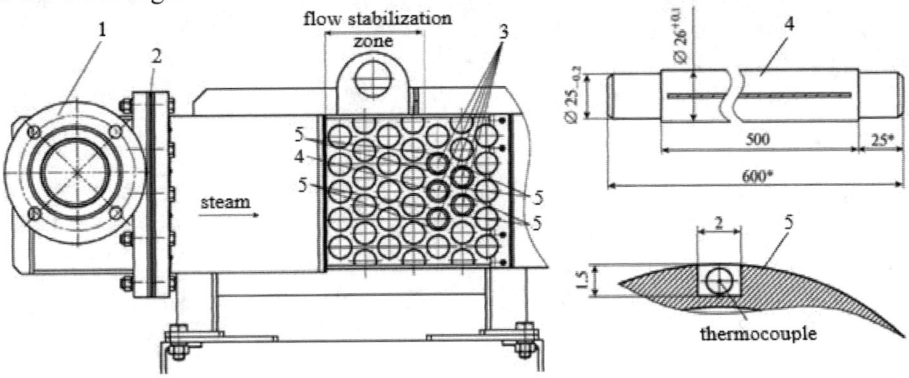

Figure 8. Working area for studying the condensation of the steam-gas mixture
1 - steam manifold, 2 - equalizing grid, 3 - measuring heat exchange tubes, 4 - copper tube without thermocouples, 5 - soldering points of thermocouples.

The steam flows through an equalizing grid. First four rows of tubes ∅26x3 used to stabilize the flow, steam condenses on the cooled tubes of 5 - 6 rows, prepared by thermocouples (4 per tube) located in grooves 1.5 mm deep. (Набор слов) The obtained experimental data almost repeat the experiments of L.D. Berman made 40 years ago [4] for pure moving steam. The test results are summarized in [5]. The condensation of steam from the SGM is well intensified by the dynamic effect of the moving mixture. To maintain a constant mixture velocity as it condenses, it is necessary to have a narrowing steam channel, the configuration of which is shown in Figure 9.

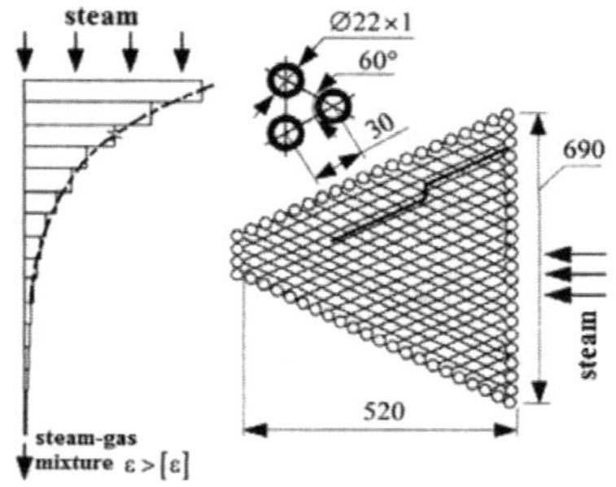

Figure 9. Channel shape of a highly efficient steam condenser for a steam-gas mixture.

Figure 10a shows a prototype of a highly efficient steam condenser at the factory bench, and Figure 10b shows the test results of this apparatus. The heat transfer coefficient at the rated load was 2.7 kW/(m^{2}K), which is comparable to an industrial condenser for almost pure steam.

a)

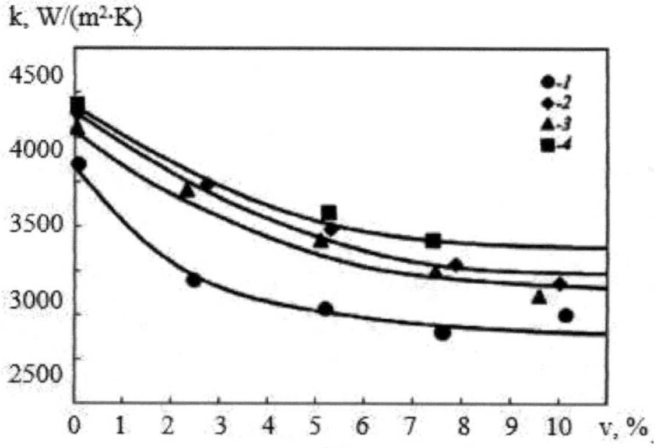

b)

Figure 10. Highly efficient condenser at the stand of PJSC "KTZ" (a) and the results of tests at various loads and gas concentraton: 1 - at a load of 0.62 q_{nom}; 2 - 0.78 q_{nom}; 3 - q_{nom}; 4 - 1.33 q_{nom} (b).

Thus, Turbocon currently has all the necessary units for creating a prototype of a gas-steam turbine plant consisting of a steam boiler, a system for additional gas steam superheating, a gas-steam turbine, and a highly efficient steam condenser from an SGM with an output of concentrated CO_2 stream from the condenser.

Acknowledgments
This work was supported by the Russian Science Foundation (grant No. 17-19-01604)

References:

[1] 2019 *Forecast of energy development in the world and in Russia* 2019, ed. A A Makarova, T A Mitrovoi and V A Kulagina (Moscow: ERI RAS – Moscow School of Management SKOLKOVO)
[2] Spielrein E E 1984 *Introduction to Hydrogen Energy* (Moscow: Ripol classic)
[3] Milman O O, Krylov V S, Ptakhin A V, Kondratev A V and Yankov G G 2017 High-efficiency condenser of steam from a steam-gas mixture *J. Therm. Eng.* **64** 874-883
[4] Fedorov V A, Milman O O and Shifrin B A 2007 *Highly efficient technology of electricity generation using organic and hydrogen fuels* (Moscow: Bauman Moscow State Technical University Press)
[5] Klimenko A V, Milman O O and Shifrin B A 2015 A high-temperature gas-and-steam turbine plant operating on combined fuel *J. Therm. Eng.* **62** 807-816
[6] *Allam power cycle.* http://en.wikipedia.org. Accessed October 25, 2021.

The investigation of the average diameter of the SiO_2 nanoparticles effect on the oil displacing efficiency from the pore in the rock formation using nanosuspension as a displacing agent

A S Lobasov[1,2], A V Minakov[1,2]

[1] Siberian Federal University, 79 Svobodny pr., Krasnoyarsk, 660041, Russia
[2] Institute of Thermophysics SB RAS, 1 Lavrentiev pr., Novosibirsk, 630090, Russia

perpetuityrs@mail.ru

Abstract. The numerical investigation of the nanofluid flow, which displaced the oil, in a microchannel was carried out. The effect of the average diameter of the SiO_2 nanoparticles on the oil displacing efficiency by nanofluids for different sizes of microchannel at various Reynolds numbers was studied. A T-shaped microchannel with a vertical channel, called a pore channel, which imitated the pore in the rock formation was considered as a computational domain. The main flow channel width and height were 200 μm. The width and height of the pore channel were varied in the range from 100 μm to 800 μm. The Reynolds number varied from 0.1 to 100. The oil recovery coefficient, which is defined as the ratio of the displacing volume of oil from the pore to the volume of the pore was considered as the main studied characteristic. The nanofluid is considered a single-phase fluid with experimentally obtained properties. The mass concentration of SiO_2 nanoparticles was 0.5%. The average diameters of nanoparticles were 5 nm, 18 nm, and 50 nm. It was found, that the oil recovery coefficient increased with a decrease in the average diameter of nanoparticles. It was obtained that the nanofluid can enhance the oil recovery several times compared to pure water.

1. Introduction

Modern oil recovery methods are considered unsatisfactory because average values of the oil recovery factor range from 25% to 40%. Therefore, it is necessary to develop enhanced oil recovery technologies. There are three main categories of such technologies available: Thermal methods, which mainly introduce heat into heavy oil reservoirs to improve the ability of heavy oil or bitumen to flow in reservoirs by changing their physical properties (viscosity and density); Gas methods, which use hydrocarbon or non-hydrocarbon gases dissolved in oil, capable of improving oil recovery by reducing the viscosity and increasing the volume of oil; Chemical methods, which mainly involve the use of long-chain molecules (polymers) to increase waterflooding efficiency or surfactants like detergents to reduce interfacial tension, which often prevents oil from being lost as it flows through the formation. However, it was shown that all these traditional methods of enhanced oil recovery have a number of serious problems, which can be solved with the use of nanoparticles with some unique useful characteristics.

Shortly, all of these enhanced oil recovery methods allow more oil to be recovered from reservoirs through various mechanisms such as interfacial tension reduction, wettability change, mobility control,

Content from this work may be used under the terms of the Creative Commons Attribution 3.0 licence. Any further distribution of this work must maintain attribution to the author(s) and the title of the work, journal citation and DOI.
Published under licence by IOP Publishing Ltd

physical properties change, and gravity drainage. However, all these traditional methods of enhanced oil recovery have a number of serious problems. For example, in gas methods, injected gas often seeps through the reservoir from injection to production wells too quickly due to the high mobility of the injected gas and oil, as a result of which a large amount of residual oil remains unrecovered from the reservoir [1, 2]. Chemical methods are often limited by the high cost of chemicals, possible formation damage, and chemical losses [3, 4]. Therefore, there is a need for less expensive, more efficient, and environmentally friendly methods of enhanced oil recovery. To solve such problems, it is possible to use nanoparticles [5], which, unlike liquids used in traditional enhanced oil recovery processes (gas, water, and chemicals), have some useful characteristics:

1. Ultra-small size. One of the main problems of chemical methods, which must be addressed in the first place, is the clogging of the pores and the locking of the injected chemicals in the porous medium. This leads to a decrease in the formation permeability and an increase in the cost of injection [6]. The most commonly used nanoparticles, such as SiO_2, TiO_2, and Al_2O_3, have sizes on the order of 1–100 nm, which is much smaller than the pore and throat sizes [7]. Thus, they can easily flow through a porous medium without significantly reducing the formation permeability or blocking the injected material, which increases the efficiency of enhanced oil recovery methods due to the injected fluids. In addition, due to the ultra-small size of nanoparticles, they are able to penetrate into some pores, where traditionally used liquids cannot get into. Thus, the nanoparticles can contact a large number of zones covered by the displacement, which makes it possible to increase the macroscopic displacement efficiency.

2. Very high surface area to volume ratio: Due to their small size, nanoparticles have a very high surface-to-volume ratio. It is clear that a large surface area increases the fraction of atoms on the surface of nanoparticles [8], thereby increasing their ability to modify various properties of displacing agents.

3. Low cost and environmental friendliness: One of the problems of using chemicals in the field is the high cost of pumping fluids. Since the price of nanoparticles is generally lower than that of chemicals, they can be widely used to enhance oil recovery in oil fields. In addition, most of the nanoparticles used are environmentally friendly materials compared to chemicals. For example, nanoparticles of silicon dioxide, which is the main component of sandstone, are often used.

Due to the aforementioned characteristics of nanoparticles, they can solve many of the existing problems faced by traditional methods of enhanced oil recovery. In this case, not the nanoparticles themselves, nanofluids, nanoemulsions, and nanocatalysts based on them are usually used.

It has been shown in many works that flooding with nanofluids makes it possible to increase oil recovery, and in different studies the magnitudes of such intensification are different: up to 10% [9, 10], up to 20% [11, 12] and up to 30% [11, 13], and work [14] shows an increase in oil recovery by 28-40%. Based on the conducted literature review, it can be concluded that the research in this area is very relevant.

It was investigated in this paper the two-phase flow of oil and nanofluid mixtures in the direct microchannels, which simulate a pore in the rock formation. The nanofluid was used as a displacing agent. The dependences of the oil recovery coefficient and the displacing efficiency on the average diameters of the nanoparticles were obtained. Two cases, when pure water was used as a displacing agent [15] and when nanofluid was used as it, were compared.

2. The computational domain and the mathematical model

The computational fluid dynamics method, namely, the numerical solution of the unsteady Navier-Stokes equations, were used:

$$\frac{\partial \rho}{\partial t}+\nabla\cdot(\rho\mathbf{u})=0, \qquad \frac{\partial \rho\mathbf{u}}{\partial t}+\nabla\cdot(\rho\mathbf{u}\mathbf{u})=-\nabla p+\nabla\tau,$$

where ρ is the fluid density, $\mathbf{u}$ and p are the fluid velocity and pressure, t is the time, $\tau=\mu(\nabla\mathbf{u})^s$ is the viscous stress tensor, μ is the fluid viscosity, and $(\nabla\mathbf{u})^s$ is the symmetric divergence-free second rank tensor. The full description of such approach can be found in many studies [15-18], so further in this paper will be given only the main points of the numerical technique. The method of finite volumes was used to find the difference analogue of convective-diffusion equations. The used calculation grids were structured and multiblock. The second-order upwind schemes were used to carrying out the approximation of transport equations convective terms. The connection between the fields of the velocity and of the pressure is realized using the SIMPLEC procedure on combined grids [19]. Such connection ensures the fulfilment of the continuity equation. To obtain the difference equations the original system was discretized. An algebraic multigrid solver was used to iterative solving those equations. To describe the two-phase interaction and flow it was used the Volume-of-Fluid method, which is detailly considered in [15]. The scheme of the computational domain is shown in figure 1. The pore in the rock formation was simulated using a T-junction microchannel with a vertical channel, called a pore channel, which imitated the pore in the rock formation, and a horizontal main flow channel. The width and height (d_{fc} and h_{fc} in figure 1, respectively) of the main channel were equal to 200 μm. The width and height (d and h in figure 1, respectively) of the pore channel were varied from 100 μm to 800 μm. The entire channel is filled with oil at the initial time moment, then it is washed out by a displacing agent. The viscosity of the oil was equal to 0.0079 Pa·s and its density was equal to and 864 kg/m^3. Pure water or nanofluid was used as the displacing agent. The viscosity of the water was equal to 1.003 mPa·s and its density was equal to 997 kg/m^3. The surface tension coefficient of the water was equal to 2.23 N/cm and the contact angle between water and oil was equal to 72°. The nanofluid was considered using the single-fluid approach. The nanofluid properties were depended on the weight concentration of SiO_2 nanoparticles of different average diameters and were taken from the experiment [20]. The parabolic velocity profile with the constant flow rate was set at the inlet of the main flow channel. The no-slip boundary conditions were set on the channel walls. The calculation grid consisted of about 2.5 million cells. This amount was enough to obtain a mesh-independent solution. The calculations were carried out in the CFD-package Ansys Fluent.

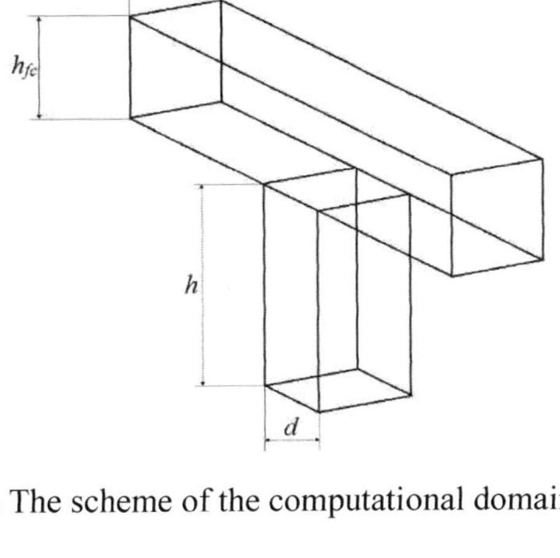

Figure 1. The scheme of the computational domain

3. Results and discussion

Numerical studies of the effect of the nanoparticles average diameters on oil displacing efficiency by the nanofluid were conducted. The Reynolds numbers varied from 0.1 to 100. The oil recovery coefficient, which defined as the ratio of the displaced volume of oil from the pore to the volume of the pore, was considered as the main investigated quantity. The results obtained for the pure water are shown in figure 2. As can be seen, pure water is a bad displacing agent, because only for pores with very huge width or very small height the oil displacing coefficient was higher than 10%. This once again confirms that oil is retained in the pores due to interfacial tension forces, and the contact angle of wetting is the main parameter affecting the efficiency of oil recovery. It is known that this angle is only 72° for water, so it cannot efficiently extract oil from narrow and deep pores. As was established in [20], for a nanofluid with a particle size of 5 nm, its value was 151°, for particles with a size of 18 nm, the contact angle was 117°, and for particles with a size of 50 nm, it was 90°. Thus, it should be expected that the oil recovery factor will decrease with an increase in the average nanoparticle size. As a result of numerical studies, it was found that this is indeed the case. It can be seen from figures 3-5, which represented the dependences of the coefficient of oil recovery by nanofluid on the Reynolds number for different pore sizes and for average nanoparticle sizes of 5 nm, 18 nm and 50 nm, respectively. It should be noted, that each configuration has been studied at the same Reynolds numbers, however, if the resulting value were lower than 0.1%, it was considered as zero. That is why some points are absent in figures 3-5.

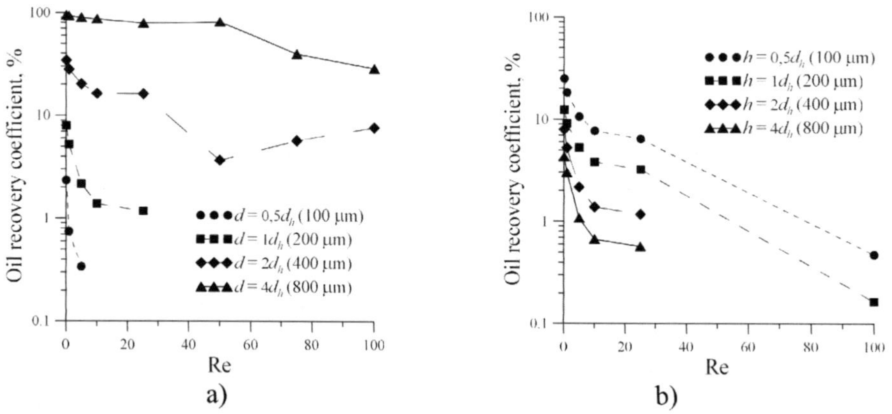

Figure 2. The dependences of the oil recovery coefficient on the Reynolds number for different widths (a) and heights (b) of pore. Pure water as displacing agent

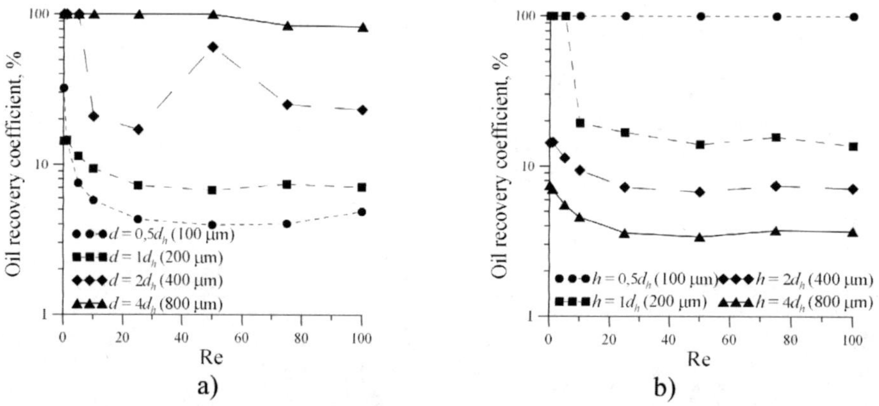

Figure 3. The dependences of the oil recovery coefficient on the Reynolds number for different widths (a) and heights (b) of pore. 0.5 wt.% nanofluid of 5 nm particles as displacing agent

The shown above tendency was confirmed by analysis of the dependences of the relative oil recovery factor for nanofluids with different average nanoparticle sizes. Namely, with an increase in the particle size, the oil recovery efficiency decreases. These dependences were obtained by dividing the nanofluid oil recovery coefficient by the pure water oil recovery coefficient. It was established that for a nanofluid with an average particle size of 18 nm, the relative oil recovery factor is greater than one for all considered geometries, except for the widest pore, and for a nanoparticles size of 50 nm, for the narrowest and widest pores. The nanofluid with the nanoparticles of 5 nm average size shown the greatest results, namely, for each considered geometry its relative oil recovery factor was greater than one. Moreover, in some cases, it reached several thousand, but it needs to be careful with the values of this quantity because such an increase is explained by the extremely low values of the oil recovery coefficient for water. For example, as can be seen from figures 2a and 3a, for a case, when the pore size was 200 μm and the Reynolds number was equal to 50, the oil recovery coefficient obtained using water as a displacing agent was almost zero, and the oil recovery coefficient obtained using nanofluid with the nanoparticles of 5 nm average size as a displacing agent was about only 8%, but it is higher than the value for water in several hundred times. Despite the lower efficiency, the values of the oil recovery factor obtained for nanofluids with larger nanoparticles can also exceed the corresponding values for pure water hundreds and thousands of times. At the same time, nanofluid based on nanoparticles with an average size of about 50 nm did not show a significant increase in the oil recovery factor in comparison with a similar value for pure water, because the values of the contact angle for such a nanofluid and pure water are quite close. In addition, it can be seen that the efficiency of oil recovery decreases with an increase in the Reynolds number. This is due to the fact that with an increase in the Reynolds number, separated flows begin to form in a channel of this configuration, whereas at low Reynolds numbers, there are no such separated flows. At even higher Reynolds numbers, such separated flows do not always fall inside the pore but pass over it. This situation takes place at small values of the channel width, therefore, e.g. for the channel width $d = 200$ μm at the Re = 50, such a dip in the value of the oil recovery factor is observed. Nanofluid significantly changes the displacement pattern, primarily due to the fact that it has a significantly higher contact wetting angle. This confirms the assumption that the main mechanism for increasing the efficiency of oil recovery due to the use of nanofluids is a significant increase in the contact angle of wetting when using the latter as a displacing agent in oil production technologies.

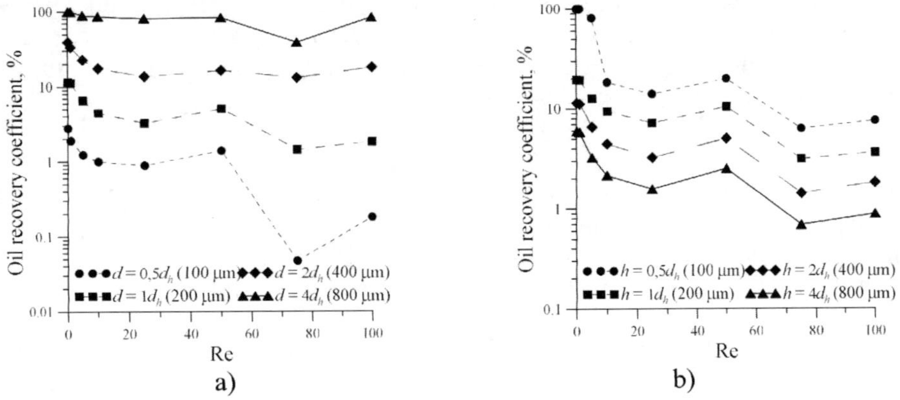

Figure 4. The dependences of the oil recovery coefficient on the Reynolds number for different widths (a) and heights (b) of pore. 0.5 wt.% nanofluid of 18 nm particles as displacing agent

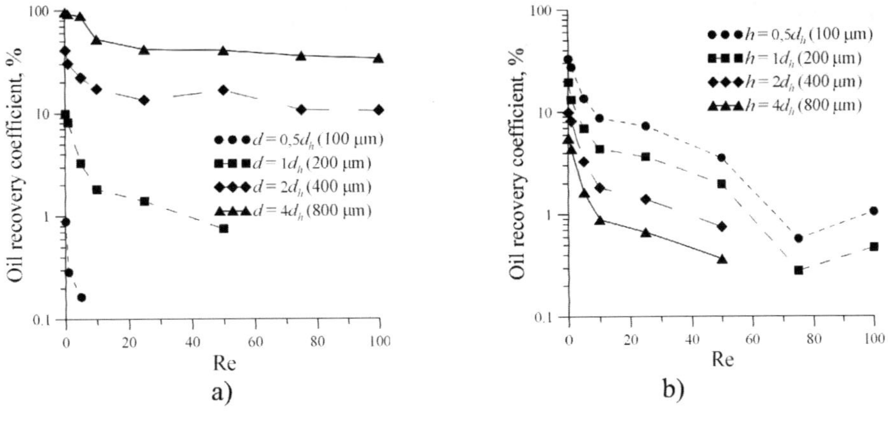

Figure 5. The dependences of the oil recovery coefficient on the Reynolds number for different widths (a) and heights (b) of pore. 0.5 wt.% nanofluid of 50 nm particles as displacing agent

4. Conclusion

The numerical studies of the effect of the average diameters of nanoparticles on the oil displacing efficiency by the water or nanofluid were carried out. It was found, that the oil recovery coefficient increased with a decrease in the nanoparticles average diameters. It was found, that the pure water can displace the oil from the pore only with huge width or very small height. The nanofluid with nanoparticles of 50 nm average diameter shown the similar results. It can be explained that the contact angle of wetting for pure water is very close to one for nanofluid. However, not only that one parameter affected the oil recovery coefficient, another important parameter is the viscosity of the displacing agent. Because the nanofluid has higher viscosity, in some cases it displaced the oil from the pore more efficient, than the pure water. The 18 nm and especially 5 nm nanofluid shows a significant increase in the efficiency of oil extraction, in some cases hundreds and thousands of times. This confirms the assumption that the main mechanism for increasing the efficiency of oil recovery due to the use of nanofluids is a significant increase in the contact angle of wetting when using the latter as a displacing agent in oil production technologies. Thus, the use of nanofluids as a displacing agent makes it possible to increase the efficiency of oil recovery, but it is necessary that such nanofluid has a high value of the contact angle of wetting.

Acknowledgments
The work is performed at partial financial support of the Russian Foundation for Basic Research, Government of Krasnoyarsk Territory, Krasnoyarsk Region Science and Technology Support Fund (Contract No. 19-48-240015: «Investigation of the oil recovery enhancement of the hydrocarbon-bearing stratum through using microfluidic technologies and by means of the nanosuspensions as the displacing agent»).

References

[1] Ahmadi Y., Eshraghi S.E., Bahrami P., Hasanbeygi M., Kazemzadeh Y., Vahedian A. 2015 *J. Petrol. Sci. Eng.* **133** 123–129
[2] Sun X., Dong M., Zhang Y., Maini B.B. 2015 *Fuel* **159** 962–973.
[3] Chang H.L., Zhang Z.Q., Wang Q.M., Xu Q.M., Guo Z.D., Sun H.Q., Cao X.L., Qiao Q. 2006 *J. Petrol. Technol.* **58** 84–89
[4] Vargo J., Turner J., Vergnani B., Pitts M.J., Wyatt K., Surkalo H., Patterson D. 2000 *SPE Reserv. Eval. Eng.* **3** 552–558
[5] Krishnamoorti R. 2006 *J. Petrol. Technol.* **58** 24–26
[6] Ahmadi M., Habibi A., Pourafshari P., Ayatollahi P. 2011 Proc. SPE Middle East Oil-Gas Show and Conf, Manama, Bahrain, 25–28 September 2011.
[7] Alomair O.A., Matar K.M., Alsaeed Y.H. Proc. SPE Asia Pacific Oil & Gas Conf. and Exhib., Adelaide, Australia, 14–16 October 2014.
[8] Zhang L.D., Liu Y.Z. 2001 *J. North China Inst. Technol.* **22** 32–43
[9] Torsater O., Li S., Hendraningrat L. 2013 *Soc. Petrol. Eng.* art. № SPE-165955-MS
[10] Haroun M.R., Alhassan S., Ansari A.A., Al Kindy N.A.M., Abou Sayed N., Abdul Kareem B.A., Sarma H.K. 2012 *Soc. Petrol. Eng.* art. № SPE-162386-MS
[11] Ogolo N.A., Olafuyi O.A., Onyekonwu M.O. 2012 *Soc. Petrol. Eng.* art. № SPE-160847-MS
[12] Roustaei, A.;Bagherzadeh, H. 2015 *J. Petrol. Explor. Prod. Technol.* **5** 27–33
[13] Kazemzadeh, Y.; Eshraghi, S.E.; Kazemi, K.; Sourani, S.; Mehrabi, M.; Ahmadi, Y. 2015 *Ind. Eng. Chem. Res.* **54** 233–239
[14] Mohebbifar, M.; Ghazanfari, M.H.; Vossoughi, M. 2015 *J. Energy Resour. Technol.* **137** 1–7
[15] Lobasov A.S., Minakov A.V. 2021 *J. Phys.: Conf. Series* **1867** art № 012028 1–6
[16] Lobasov A.S., Minakov A.V. Rudyak V.Ya. 2018 *J. Eng. Phys.& Thermophys.* **91(1)** 124–35
[17] Lobasov A.S., Minakov A.V. 2018 *Chem. Eng.& Proc.: Proc. Intens.* **124** 11–23
[18] Lobasov A.S., Minakov A.V. Rudyak V.Ya. Shebeleva A.A. Kuznecov V.V. 2018 *Chem. Eng.& Proc.: Proc. Intens.* **134** 105–14
[19] Patankar S. 1980 *Numerical heat transfer and fluid flow* (New York: Hemisphere Publishing Corporation/McGraw Hill Book Company) p 197
[20] Minakov A.V., Pryazhnikov M.I., Suleymana Y.N., Meshkova V.D., Guzei D.V. 2021 *J. Molec. Liq.* **327** 1–10

Composition of the gas-plasma phase in the radioactive graphite - water vapor system

N M Barbin[1,2], A M Kobelev[1], D I Terent'ev[3] and S A Titov[1]

[1]Research department (as part of the educational and scientific complex for firefighting and rescue operations), Ural Institute of the State Fire Service EMERCOM of Russia, 22 Mira Street, Ekaterinburg 620062, Russia
[2]Institute of Physicochemical Problems and Technosphere Safety, Ural State Agrarian University, 42 Karla Libknekhta Street, Ekaterinburg 620075, Russia
[3]Research department (as part of the educational and scientific complex for ensuring fire safety of objects and settlements), Ural Institute of the State Fire Service EMERCOM of Russia, 22 Mira Street, Ekaterinburg 620062, Russia

e-mail: nmbarbin@mail.ru

Abstract. In this work, the composition and thermophysical properties of the "Reactor graphite-H_2O" system at temperatures from 2123 to 3223 K are calculated. It was found that the main components of the vapor phase at a temperature of 2123-2923 K: carbon dioxide, carbon monoxide, water vapor, hydroxide, hydrogen, atomic hydrogen. At temperatures above 3223 K, oxygen and atomic oxygen are added to the gases present. The balances of uranium and plutonium are considered. Uranium at temperatures above 2123 K is present in the system in the form of gaseous and ionized uranium dioxide and trioxide. Plutonium at temperatures above 2123 K is present in the system in the form of gaseous and ionized plutonium oxide, gaseous plutonium dioxide. The calculation of thermophysical properties for the considered system is carried out.

1. Introduction

Reactor graphite in the form of blocks is used as a neutron reflector in nuclear reactors of various generations. Various types of structural elements of technological equipment are also made from reactor graphite: graphite rings of contact between the process channel and the graphite block; graphite bushings for fuel assemblies; displacers of rods of the control and protection system, etc [1].

In addition to thermal, chemical and mechanical loads, reactor graphite is exposed to powerful radioactive radiation during operation, which leads to its degradation. The activity of graphite in the reactor core is due to: neutron activation of carbon in the graphite matrix; contamination of graphite with products of activation of the gas filling the masonry; hitting the masonry with coolant, products of erosion and corrosion; and fission products and nuclear fuel getting into masonry during a "wet" accident. The level of activation by neutrons depends on its chemical composition graphite, duration of irradiation, flux and spectrum of neutrons. The activated impurities in gases are ^{3}H, ^{14}C, ^{36}Cl, ^{41}Ar, ^{13}N, ^{16}N. Radioactive products of corrosion and erosion that contaminate graphite in their mutual contact with structural elements (channels, rods, supports, etc.) are mainly radioisotopes of zirconium and

Content from this work may be used under the terms of the Creative Commons Attribution 3.0 licence. Any further distribution of this work must maintain attribution to the author(s) and the title of the work, journal citation and DOI.
Published under licence by IOP Publishing Ltd

Journal of Physics: Conference Series **2150** (2022) 012026 doi:10.1088/1742-6596/2150/1/012026

niobium, the concentration of which is low. Contamination of graphite with a fuel composition and fission products during "wet" accidents occurs due to the ingress of gamma-emitting nuclides into the graphite stack: ^{134}Cs, ^{137}Cs, ^{60}Co, ^{55}Fe, ^{154}Eu, ^{155}Eu, ^{151}Sm, ^{244}Cm, ^{241}Am, ^{243}Am, ^{239}Pu, ^{106}Ru, ^{90}Sr. Most often, such contamination is observed on the surface of graphite blocks [2].

The largest amount of radioactive graphite is found in UK (eighty-six thousand tons), the USA (fifty-five thousand tons), the Russian Federation (sixty thousand tons), France (twenty-three thousand tons). A smaller part of radioactive graphite is located in Ukraine (5.7 thousand tons), Lithuania (3.8 thousand tons), Spain (3.7 thousand tons), North Korea (3.5 thousand tons), Italy (three thousand tons), Japan (three thousand tons). tons), Belgium (2.5 thousand tons), Germany (two thousand tons) [2].

For most of the nuclear reactors included in the decommissioning programs, a delayed decision was adopted, which provides for mothballing and controlled maintenance for a long time. At the moment, no final decision on the problem of using radioactive graphite has been made in the world. There are the following options for the processing of reactor graphite are considered: heat treatment (up to 1400 °C), high-temperature oxidation, acid treatment, washing of graphite with synthetic surfactants, in molten salts, the use of spark erosion, plasmification, reuse of deactivated graphite, processing in boiling bed, gasification processing, gasification of graphite using superheated steam [2].

The aim of the work is to calculate the composition and thermophysical properties of the vapor phase for the "Reactor graphite-H_2O" system. The task of the work is to conduct a thermodynamic analysis of the "Reactor graphite-H_2O" system at temperatures from 2123 to 3223 K.

2. Calculation Method

Exploring System Behavior "Reactor graphite-H_2O" is studied using the method of thermodynamic modeling. The thermodynamic modeling method is a computational experiment that allows you to analyze states and processes and draw conclusions about the behavior of the objects under study based on model concepts. It follows from the laws of thermodynamics that in a state of equilibrium the entropy of an isolated system is maximum. Therefore, the problem of thermodynamic calculation is reduced to finding the coordinates of the conditional maximum of entropy. Thermodynamic modeling was performed in the TERRA software package. TERRA allows for thermodynamic calculations of the phase composition of arbitrary heterogeneous systems, as well as their thermodynamic and transport properties. To carry out thermodynamic calculations, thermodynamic databases were used: IVTANTHERMO, HSC Chemistry, etc [3 – 9].

To carry out thermodynamic modeling of the "Reactor graphite-H_2O" system, the following initial composition was selected: condensed phase – 25.003 % (C – 24.999 %, Eu – $1.999\cdot10^{-6}$ %, Am – $2.499\cdot10^{-6}$ %, Sr – $2.499\cdot10^{-6}$ %, Cs – $9.999\cdot10^{-7}$ %, Ni – $1.999\cdot10^{-6}$ %, Pu – $1.799\cdot10^{-5}$ %, Ca – $6.749\cdot10^{-5}$ %, Cl – 0.0004 %, U – 0.002 %); vapor phase (H_2O - 74.997%). The investigated temperature range: 2123-3223 K, with a step of 100 K. Constant pressure is 0.1 MPa.

3. Results and discussion

Table 1 shows composition of the vapor phase of the "Reactor graphite-H_2O" system at temperatures from 2123 to 3223 K.

Table 1. Composition of the vapor phase (atm.) of the "Reactor graphite-H_2O" system at temperatures from 2123 to 3223 K

T (K)	CO_2	CO	H_2O	OH	H_2	H	O_2	O
2123	0.044	0.2889	0.2887	0.0002	0.3759	0.0023	$3.4\cdot10^{-7}$	$1.05\cdot10^{-6}$
2223	0.042	0.2905	0.2901	0.0005	0.3728	0.004	$1.2\cdot10^{-6}$	$3.75\cdot10^{-6}$
2323	0.0403	0.2917	0.2908	0.0009	0.3694	0.0068	$3.9\cdot10^{-6}$	$1.2\cdot10^{-5}$

2423	0.0388	0.2924	0.2907	0.0017	0.3654	0.011	$1.2\cdot10^{-5}$	$3.53\cdot10^{-5}$
2523	0.0374	0.2926	0.2895	0.0029	0.3604	0.017	$3.1\cdot10^{-5}$	$9.52\cdot10^{-5}$
2623	0.0361	0.2921	0.2868	0.0049	0.3543	0.0255	$7.8\cdot10^{-5}$	0.0002
2723	0.0347	0.2908	0.2821	0.0079	0.3467	0.037	0.0002	0.0006
2823	0.0333	0.2887	0.2746	0.0122	0.3374	0.0522	0.0004	0.0012
2923	0.0317	0.2856	0.2635	0.018	0.3263	0.0716	0.0008	0.0025
3023	0.0297	0.2812	0.2479	0.0255	0.3134	0.0958	0.0015	0.0049
3123	0.0273	0.2756	0.2272	0.0344	0.2988	0.1253	0.0026	0.0089
3223	0.0243	0.2685	0.2012	0.0439	0.2825	0.1602	0.0041	0.0152

The pressure of carbon monoxide rising from 0.2888 to 0.29242 atm., water vapor from 0.28869 to 0.290741 atm., atomic hydrogen from 0.00228 to 0.01097 atm., while the pressure of hydrogen is dropping from 0.37588 to 0.36537 atm., carbon dioxide from 0.04401 to 0.03877 atm. with an increase in temperature from 2123 to 2423 K. The pressure of atomic hydrogen rising from 0.01097 to 0.07158 atm., hydroxide from 0.00166 to 0.018 atm., while the pressure of carbon monoxide dropping from 0.29242 to 0.28557 atm., water vapor from 0.29074 to 0.26349 atm., hydrogen from 0.36537 to 0.32631 atm., carbon dioxide from 0.03877 to 0.0317 atm., with an increase in temperature from 2423 to 2923 K. The pressure of atomic hydrogen rising from 0.07158 to 0.16021 atm., hydroxide from 0.018 to 0.04394 atm., atomic oxygen from 0.00251 to 0.01515 atm., oxygen from 0.00079 to 0.00413 atm., while the pressure of carbon monoxide dropping from 0.28557 to 0.26853 atm., water vapor from 0.26349 to 0.202 atm., hydrogen from 0.32631 to 0.28248 atm., carbon dioxide from 0.0317 to 0.02432 atm., with an increase in temperature from 2923 to 3223 K.
Table 2 shows the distribution of uranium and plutonium in vapor phase of the "Reactor graphite-H_2O" system at temperatures from 2123 to 3223 K.

Table 2. Distribution of uranium and plutonium in vapor phase (mol%) of the "Reactor graphite-H_2O" system at temperatures from 2123 to 3223 K

T (K)	UO_2	UO_3	UO_2^+	UO_3^-	PuO	PuO_2	PuO^+
2123	4.0489	75.5938	8.2983	12.0589	1.9303	97.046	1.0237
2223	4.7136	72.89	8.901	13.4951	2.3881	96.3252	1.2866
2323	5.4069	70.2571	9.4387	14.8968	2.8967	95.5173	1.5859
2423	6.12	67.6897	9.9749	16.2143	3.4534	94.6227	1.9238
2523	6.8465	65.1651	10.5462	17.4403	4.0555	93.6411	2.3033

2623	7.5915	62.7388	11.1198	18.5467	4.7021	92.5687	2.7289
2723	8.3744	60.3963	11.7346	19.4893	5.396	91.3931	3.2104
2823	9.1908	58.0992	12.4733	20.228	6.1469	90.0822	3.7702
2923	10.0602	55.8854	13.4273	20.6131	6.976	88.5648	4.4581
3023	11.0813	53.6005	14.8571	20.4388	7.9144	86.7089	5.375
3123	12.2698	51.0899	17.1941	19.4101	9.0138	84.2556	6.7279
3223	13.6674	47.9796	20.9731	17.3203	10.3271	80.7887	8.8797

Let us consider the distribution of uranium in the vapor phase. Ionized uranium trioxide reaches 20.6131 mol%, ionized uranium dioxide - 10.0601 mol%, vaporous uranium dioxide - 10.0602 mol%, at the same time, vaporous uranium trioxide decreases to 55.8854 mol% with an increase in temperature from 2123 to 2923 K. Ionized uranium dioxide reaches 20.9731 mol%, vaporous uranium dioxide to 13.6674 mol%, at the same time, vaporous uranium trioxide decreases to 47.9796 mol%, ionized uranium trioxide - 17.3203 mol% with an increase in temperature from 2923 to 3223 K.

Let us examine the distribution of plutonium in the vapor phase. Vaporous plutonium oxide reaches 10,3271 mol%, ionized plutonium oxide to 8.8797 mol%, at the same time vaporous plutonium dioxide decreases to 80.7887 mol% with an increase in temperature from 2123 to 3223 K.

Table 3 shows the thermophysical characteristics of the "Reactor graphite-H_2O" system at temperatures from 2123 to 3223 K.

Table 3. Thermophysical characteristics of the "Reactor graphite-H_2O" system at temperatures from 2123 to 3223 K

T (K)	Number of moles (mol/kg)	Internal energy (kj/kg)	Entropy (kj/(kg K))	Enthalpy (kj/kg)	Volume (m^3/kg)
2123	62.5223	-4199.295	15.5571	-3250.665	11.03615
2223	62.585	-3959.57	15.69175	-2957.945	11.5676
2323	62.68625	-3700.45	15.8293	-2645.075	12.10765
2423	62.84375	-3413.765	15.97315	-2303.47	12.66075
2523	63.0809	-3088.6	16.1274	-1921.65	13.23315
2623	63.42835	-2710.555	16.2971	-1484.42	13.83365
2723	63.9263	-2260.715	16.48855	-971.763	14.47415

2823	64.62735	-1714.285	16.7097	-357.415	15.1707
2923	65.5997	-1039.175	16.9703	392.712	15.94495
3023	66.9294	-195.9615	17.28175	1320.674	16.8254
3123	68.7178	862.4935	17.6572	2476.885	17.84735
3223	71.06905	2183.095	18.10865	3911.925	19.05005

The number of moles increases to 63.0809 mol/kg with an increase in temperature from 2123 to 2523 K according to a linear law. The number of moles increases more intensively to 71.06905 mol/kg (14.3 times) with an increase in temperature from 2523 to 3223 K.

Internal energy increases to -3088.6 kJ/kg with an increase in temperature from 2123 to 2523 K. Internal energy increases more intensively to 2183.095 kJ/kg (4.7 times) with an increase in temperature from 2523 to 3223 K.

Entropy increases to 16.1274 kj/(kg K) with an increase in temperature from 2123 to 2523 K. Entropy increases more intensively to 18.10865 kj/(kg K) (3.4 times) with an increase in temperature from 2523 to 3223 K.

Enthalpy increases to -1921.65 kj/kg with an increase in temperature from 2123 to 2523 K. Enthalpy increases more intensively to 3911.925 kj/kg (4.3 times) with an increase in temperature from 2523 to 3223 K.

Volume increases to 13.23315 m^3/kg with an increase in temperature from 2123 to 2523 K according to a linear law. Volume increases more intensively to 19.05005 m^3/kg with an increase in temperature from 2523 to 3223 K.

4. Conclusion

The work carried out a thermodynamic study of the composition and thermophysical properties in vapor phase of the "Reactor graphite-H_2O" system.

The results of thermodynamic calculations are presented in the following tables: Composition of the vapor phase of the "Reactor graphite-H_2O" system at temperatures from 2123 to 3223 K; Distribution of uranium and plutonium in vapor phase of the "Reactor graphite-H_2O" system at temperatures from 2123 to 3223 K; Thermophysical characteristics of the "Reactor graphite-H_2O" system at temperatures from 2123 to 3223 K.

Analyzing the tables, the following conclusions can be drawn: at $T > 2923$ K, the main components of the vapor phase for the system under consideration are: carbon dioxide, carbon monoxide, water vapor, hydroxide, hydrogen, atomic hydrogen, oxygen, atomic oxygen; uranium at temperatures above 2123 K is present in the system in the form of ionized and vaporous uranium dioxide, vaporous and ionized uranium trioxide; plutonium at temperatures above 2123 K is present in the system in the form of ionized and vaporous plutonium oxide, vaporous plutonium dioxide; at T = 2123-2523 K, the thermophysical properties of the vapor phase of the system under consideration change linearly, and at temperatures above 2523 K, they change significantly.

References

[1] Marsden B J, Jones A N, Hall G N, Treifi M and Mummery P M 2016 *Structural Materials for Generation IV Nuclear Reactors* (Cambridge: Woodhead publishing) p 495

[2] Dorofeev A N, Komarov E A, Zaharova E V, Volkova A G, Linge I I, Ivanov A Yu, Utkin S S, Pavlyuk A O and Kotlyarevskij S G 2019 *Radioactive waste* **2** 18

[3] Prigozhin I, Defay R 2009 *Chemical thermodynamics* (Moscow: Binom Publishing) p 27

[4] Kirillin V A, Sychev V V and Shejndlin A E 1983 *Technical thermodynamics* (Moscow: Energoatomizdat) p 6
[5] Fihtengol'c G M 2017 *Differential and integral calculus course. Vol. 1.* (Saint Petersburg: Publishing house Lan) p 93
[6] Gurvich L V, Vejc I V, Medvedev V A and etc. 1978 *Thermodynamic properties of individual substances. Reference edition: in 4 volumes* (Moscow: Publishing House Science Year) p 20
[7] Barbin N M, Kobelev A M, Terent'ev D I and Alekseev S G 2020 *Journal of Physics: Conference Series* **2020** 012066
[8] Barbin N M, Sidash I A, Terent'ev D I and Alekseev S G 2017 *Journal of Physics: Conference Series* **2017** 012320
[9] Barbin N M, Sidash I A, Terent'ev D I and Alekseev S G 2020 *Russian metallurgy (Metally)* **8** 743

Mathematical modelling of thermal microexplosion in a catalyst granule with point sources of heat release

I V Derevich and A Yu Fokina

Moscow State Technical University by N.E. Bauman (BMSTU), 2-ya Baumanskaya Str. 5 (main Building), Moscow, Russian Federation 105005

DerevichIgor@bmstu.ru

Abstract. A mathematical model of heat and mass transfer in a spherical catalyst granule is proposed. Exothermic synthesis reactions are carried out on point active centres located inside a porous ceramic granule. From the surface of the granule the heat of catalytic reactions is removed into liquid synthesis products. The rate of a chemical reaction is modelled by a modified Arrhenius law. In contrast to the homogeneous model of a catalytic granule methods for calculating heat transfer processes in a system of point, active centres do not develop. An iterative procedure is suggested to calculate the unknown temperature and concentration of the reagent at the active centre. It is shown that the temperature of the active centres is significantly higher than in the volume of the granule. The results of modelling a thermal explosion with increasing granule size and reactor temperature are presented.

1. Introduction

Modern microtechnologies allow producing porous ceramic granules of catalysts with point active centers of the synthesis reaction [1, 2]. The synthesis processes are exothermic [3, 4]. The active centers on which the synthesis reactions occur are located in the volume for the thermal boundary layers around the microparticles not to overlap. In this case, the homogeneous model of the catalytic granule is incorrect [4 – 6]. Modeling of heat and mass transfer processes in a granule is complicated by the fact that the values of the temperature and concentration of the reagent are unknown. The paper proposes a self-consistent procedure for calculating the temperature and concentration at the active centers and in the volume of the granule. It is shown that when the thermal stability of the catalyst granule is lost, the temperature of the active centers is significantly higher than in the volume of the granule. Thermal microexplosion can lead to the destruction of the point centers of catalysis without visible damage to the catalyst granule. The thermal explosion of the granule as a whole, which leads to its destruction, is also investigated.

2. Problem statement

2.1. Basic equations

We consider a spherical catalyst granule with active microparticles on which synthesis reactions are taking place. The equation for the temperature of a granule with point active centers of heat release has the form

Content from this work may be used under the terms of the Creative Commons Attribution 3.0 licence. Any further distribution of this work must maintain attribution to the author(s) and the title of the work, journal citation and DOI.
Published under licence by IOP Publishing Ltd

$$\rho_{gr}c_{gr}\frac{\partial T_{gr}}{\partial t}=\lambda_{gr}\Delta T_{gr}+\sum_{i=1}^{N_0}Q_0\Phi_0\left(T_0^{(i)},C_0^{(i)}\right)\delta\left(\mathbf{r}-\mathbf{r}_0^{(i)}\right)$$

Here ρ_{gr}, c_{gr}, λ_{gr} are the density, heat capacity and heat conduction of the granule; Q_0 is the heat of exothermic synthesis reaction; $\mathbf{r}_0^{(i)}$ are the position vectors of the i-th active centers; N_0 is the number of active centers in the granule; $\Phi_0(T,C)$ is the semiempirical function describing the rate of consumption of gaseous reagent; $T_0^{(i)}$, $C_0^{(i)}$ are the temperature and concentration of reagent at the i-th active center; $\delta(\mathbf{r})=\delta(x)\delta(y)\delta(z)$ is Dirac delta-function, which is normalized as

$$\int\delta(\mathbf{r})\mathbf{dr}=1$$

Here $\mathbf{dr}=dxdydz$ is the differential of the volume.

The boundary condition on the surface of the granule has the form

$$-\lambda_{gr}\left.\frac{\partial T_{gr}}{\partial r}\right|_{|\mathbf{r}|=d_{gr}/2}=\alpha_0\left(T_{gr}\big|_{|\mathbf{r}|=d_{gr}/2}-T_R\right)$$

Here d_{gr} is the diameter of the granule; T_R is the temperature in the reactor tube outside the granule; α_0 is the heat transfer coefficient between the granule surface and liquid products.

The equation for the concentration of gaseous regents C inside the granule is written as

$$\frac{\partial C}{\partial t}=D_C\Delta C-\sum_{i=1}^{N_0}\Phi_0\left(T_0^{(i)},C_0^{(i)}\right)\delta\left(\mathbf{r}-\mathbf{r}_0^{(i)}\right)$$

Here D_C is the effective diffusion coefficient of the reagent inside the porous granule.

The boundary condition describes the mass exchange by reagent through the granule surface

$$-D_C\left.\frac{\partial C}{\partial r}\right|_{r=d_{gr}/2}=\beta_C\left(C\big|_{r=d_{gr}/2}-C_R\right)$$

Here C_R is the concentration of reagent outside the granule; and β_C is the mass transfer coefficient.

2.2. *Equations in dimensionless form*

We rewrite the equations of heat balance, concentration and boundary conditions in a dimensionless form. We define the dimensionless temperature and concentration as $\Theta=(T_{gr}-T_R)/T_R$, and $\Psi_C=(C_R-C)/C_R$. The dimensionless time is $\xi=t\chi_{gr}/(d_{gr}/2)^2$ ($\chi_{gr}=\lambda_{gr}/(\rho_{gr}c_{gr})$ is the thermal diffusivity of the granule); dimensionless coordinate is $\eta=2r/d_{gr}$.

The equation and the boundary condition for the granule temperature take dimensionless form

$$\frac{\partial\Theta}{\partial\xi}=\Delta\Theta+4\pi\sum_{i=1}^{N_0}\Xi_\Theta^{(i)}\delta\left(\boldsymbol{\eta}-\boldsymbol{\eta}_0^{(i)}\right)\qquad\left.\left(\frac{\partial\Theta}{\partial\eta}+\alpha_0^*\Theta\right)\right|_{|\boldsymbol{\eta}|=1}=0\tag{1}$$

Here $\alpha_0^*=\alpha_0(d_{gr}/2)/\lambda_{gr}$ is the dimensionless coefficient of heat transfer; and $\Xi_\Theta^{(i)}$ is the dimensionless heat release during the synthesis reactions (see, for example, [3])

$$4\pi\Xi_\Theta^{(i)}=\Omega_\Theta^{(i)}\Gamma\left(\Theta_0^{(i)}\right)\qquad\Gamma(\Theta)=\exp\left[E_a^*\Theta/(1+\Theta)\right]\left\{1+b_0\exp\left[E_b^*\Theta/(1+\Theta)\right]\right\}^{-2}$$

Here $E_a^*=E_a/(R_U T_R)$, $E_b^*=E_b/(R_U T_R)$ are the dimensionless activation energy; R_U is the universal gas constant; $\Omega_\Theta^{(i)}=Q_0a_0\exp(-E_a^*)/\lambda_{gr}T_R(d_{gr}/2)$ is the dimensionless thermal effect of the reaction; and a_0, b_0 is the empirical function that depends on the concentration of the reagent.

Equation and boundary condition for the dimensionless concentration are

$$\mathrm{Le_C}\frac{\partial\Psi_C}{\partial\xi}=\Delta\Psi_C+4\pi\sum_{i=1}^{N_0}\Xi_C^{(i)}\delta\left(\boldsymbol{\eta}-\boldsymbol{\eta}_0^{(i)}\right)\qquad\left\{\frac{\partial\Psi_{CO}}{\partial\eta}+\beta_{CO}^{*}\Psi_{CO}\right\}\Bigg|_{|\boldsymbol{\eta}|=1}=0$$

Here $\beta_C^*=\beta_C\left(d_p/2\right)/D_C$ is the dimensionless coefficient of mass transfer; $\mathrm{Le_C}=\chi_{gr}/D_C$ is the Lewis number; and the dimensionless rate of reagent consumption is equal to

$$4\pi\Xi_C^{(i)}=\Omega_C^{(i)}\Gamma\left(\Theta_0^{(i)}\right)\qquad\Omega_C^{(i)}=a_0\exp\left(-E_a^*\right)/D_C C_R\left(d_{gr}/2\right)$$

3. Stationary distribution of temperature and reagent concentration in the granule

The loss of the thermal stability of the granule occurs as a result of random fluctuations near the stationary state. The temperature and concentration of the reagent on the active centers are unknown. The calculation of these parameters is carried out in a self-consistent manner. At the first stage, we find an analytical solution for a given power of heat release at the point centers of synthesis. At the next stage, we perform an iterative refinement of the temperature and concentration of the reagent at the active centers.

3.1. The solution for the specified heat releases on active centres

We describe the solution for temperature, because for concentration the calculations are carried out similarly. The stationary temperature distribution is the solution of the following problem

$$\Delta\Theta+4\pi\sum_{i=1}^{N_0}\Xi_\Theta^{(i)}\delta\left(\boldsymbol{\eta}-\boldsymbol{\eta}_0^{(i)}\right)=0\qquad\left(\partial\Theta/\partial\eta+\alpha_0^*\Theta\right)\Big|_{|\boldsymbol{\eta}|=1}=0\tag{2}$$

At the first stage, we set the heat generation capacity $\Xi_\Theta^{(i)}=\mathrm{const}$. The temperature distribution inside the granule (Equation 2) consists of two components

$$\Theta(\boldsymbol{\eta})=\Theta'(\boldsymbol{\eta})+\Theta''(\boldsymbol{\eta})$$

The $\Theta'(\boldsymbol{\eta})$ component describes the solution of an inhomogeneous Equation 2 in an infinite medium. The component $\Theta''(\boldsymbol{\eta})$ is the solution of the Laplace equation in a spherical coordinate system. The stationary temperature distribution inside the granule has the form

$$\Theta(\boldsymbol{\eta})=\sum_{i=1}^{N_0}\Xi_\Theta^{(i)}\left\{\frac{1}{\left|\boldsymbol{\eta}-\boldsymbol{\eta}_0^{(i)}\right|}+\sum_{n=0}^{\infty}\sum_{m=0}^{n}C_{m,n}^{(i)}\eta^n\cos\left[m\left(\varphi-\varphi_0^{(i)}\right)\right]P_n^m(\cos\theta)\right\}\tag{3}$$

Here $P_n^m(x)$ are the associated Legendre polynomials [7]; the coefficients $C_{m,n}^{(i)}$ are found from the boundary condition in Equation 2.

3.2. Iterative calculation of temperatures and concentrations at active centres

The next step in the self-consistent method consists in averaging the obtained temperature distribution in the granule over the volume of the active centre. We perform the integration of the granule temperature (Equation 3) over the volume of the selected active center

$$\Theta_0^{(j)}=\frac{1}{V_0}\int_{V_0}\Theta\left(\boldsymbol{\eta}_0^{(j)}+\boldsymbol{\eta}_1\right)d\boldsymbol{\eta}_1$$

Here V_0 is the volume of the active reaction center; $d\boldsymbol{\eta}'$ is the differential of the volume of the center. As a result of calculations, we obtain a system of transcendental algebraic equations for self-consistent calculation of the temperature of active centers

$$\Theta_0^{(j)}=\sum_{\substack{i=1\\i\neq j}}^{N_0}\left\{\frac{\Xi_\Theta^{(i)}}{\left|\boldsymbol{\eta}_0^{(j)}-\boldsymbol{\eta}_0^{(i)}\right|}\right\}+3\frac{\Xi_\Theta^{(j)}}{\delta_0}+\sum_{i=1}^{N_0}\Xi_\Theta^{(i)}\left\{\sum_{n=0}^{\infty}\sum_{m=0}^{n}C_{m,n}^{(i)}\cos\left(m\left(\varphi_0^{(j)}-\varphi_0^{(i)}\right)\right)P_n^m\left(\cos\left(\theta_0^{(j)}\right)\right)\left(\eta_0^{(j)}\right)^n\right\}\tag{4}$$

Equation 4 takes into account both the boundary conditions on the surface of the granule and the mutual influence of all active centers inside the granule. The solution of the equation for the concentration of the reagent inside the granule is found in a similar way. A system of equations for the concentration of the reagent on the active centers is obtained in a self-consistent manner. For the iterative solution of Equation 4, we use an effective method of finding a stable solution by solving a system of ordinary differential equations [8]

$$\tau_0 \frac{d\Theta_0^{(j)}}{dt} = \sum_{\substack{i=1 \\ i\neq j}}^{N_0} \frac{\Xi_\Theta^{(i)}}{\left|\boldsymbol{\eta}_0^{(j)} - \boldsymbol{\eta}_0^{(i)}\right|} + \frac{3\Xi_\Theta^{(j)}}{\delta_0} + \sum_{i=1}^{N_0} \Xi_\Theta^{(i)} \sum_{n=0}^{\infty} \sum_{m=0}^{n} C_{m,n}^{(i)} \cos\left(m\left(\varphi_0^{(j)} - \varphi_0^{(i)}\right)\right) P_n^m\left(\cos\left(\theta_0^{(j)}\right)\right)\left(\eta_0^{(j)}\right)^n - \Theta_0^{(j)}$$

Here τ_0 is the relaxation parameter selected from the stability condition of the solution.

The stationary solution of the system of equations $d\Theta_0^{(j)}/dt = 0$ presented above leads to the desired values of the temperature of the active reaction centers (see Equation 4).

3.3. The average temperature of the granule with localized reaction centres

The thermal explosion of the granule as a whole is investigated on the basis of the dynamics of changes in the average temperature of the granule

$$\bar{\Theta}(\xi) = \frac{1}{V_{gr}^*} \int_{V_{gr}^*} \Theta(\xi, \boldsymbol{\eta})\, d\boldsymbol{\eta}$$

Here V_{gr}^* is the volume of the granule in dimensionless variables.

We integrate Equation 1 by the volume of the granule

$$\frac{d\bar{\Theta}}{d\xi} = \frac{1}{V_{gr}^*} \int_{V_{gr}^*} \Delta\Theta\, d\boldsymbol{\eta} + \frac{4\pi}{V_{gr}^*} \sum_{i=1}^{N_0} \Xi_\Theta^{(i)}$$

Using the Ostrogradsky - Gauss theorem, we obtain

$$\frac{d\bar{\Theta}}{d\xi} = -\alpha_0^* \frac{S_{gr}^*}{V_{gr}^*} \tilde{\Theta} + \frac{4\pi}{V_{gr}^*} \sum_{i=1}^{N_0} \Xi_\Theta^{(i)}$$

Here $\tilde{\Theta}(\xi)$ is the temperature averaged over the granules surface area S_{gr}^*

$$\tilde{\Theta}(\xi) = \frac{1}{S_{gr}^*} \oint_{S_{gr}^*} \Theta(\boldsymbol{\eta}, \xi)\big|_{\boldsymbol{\eta}\in S_{gr}^*}\, dS_{gr}^*$$

For a spherical granule we have the equation for the average temperature of the granule

$$\frac{d\bar{\Theta}}{d\xi} = -3\alpha_0^* \tilde{\Theta} + 3\sum_{i=1}^{N_0} \Xi_\Theta^{(i)}$$

The mean temperature of the granule surface and mean temperature of active centers are calculated by the formulas

$$\tilde{T}_{gr} = \left(1 + \tilde{\Theta}\right) T_R \qquad \bar{T}_0 = \left(1 + \bar{\Theta}_0\right) T_R$$

4. Calculation results

The rate constants of chemical reactions and the thermal effect correspond to the synthesis of artificial oil by the Fischer-Tropsch method [9, 10]. An increase in the reactor temperature leads to a thermal explosion. There are two scenarios that can be distinguished. Microthermal explosion inside the granule and loss of thermal stability of the granule as a whole. Figure 1 shows the overheating of the active centers in relation to the average temperature of the granule.

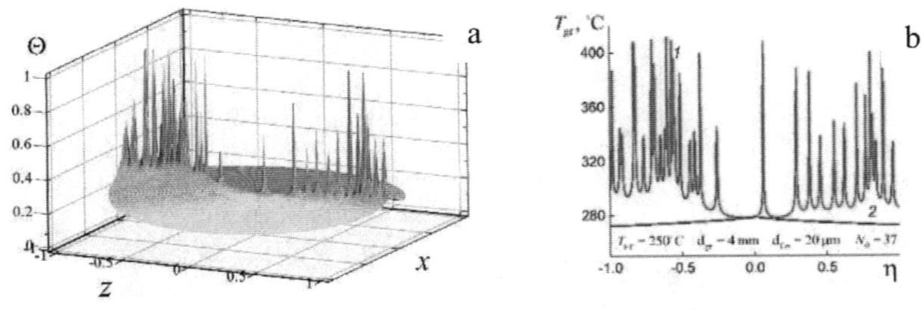

Figure 1. Temperature distribution inside the catalyst granule when the active centers are located on the same diameter (a). The average temperature of the granule is shown in blue line (b).

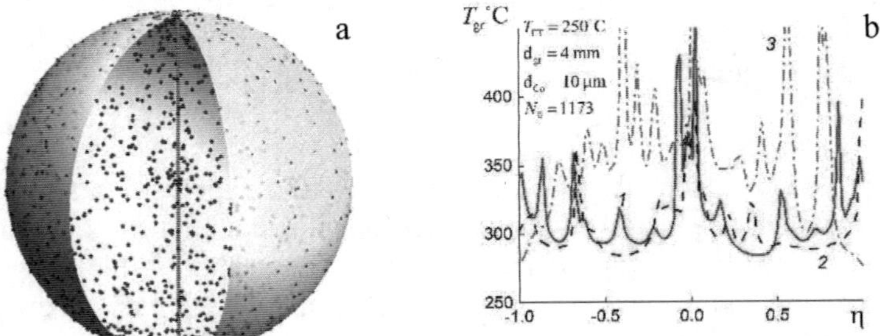

Figure 2. Distribution of active centers in the granule volume (a) and temperature distribution in the volume along three randomly selected diameters (b).

It can be seen that during thermal microexplosion, the temperatures of the active centers significantly exceed the surface temperature of the granule. Significant overheating can lead to the destruction of the reaction centers. At the same time, the surface temperature of the granule does not reflect the real thermal situation inside the granule with point active centers. Modern microtechnologies allow us to place active centers inside the granule according to a certain law. Figure 2 shows a granule with random uniformly distributed point centers of the reaction. In this case, the temperature distribution inside the granule is significantly inhomogeneous either.

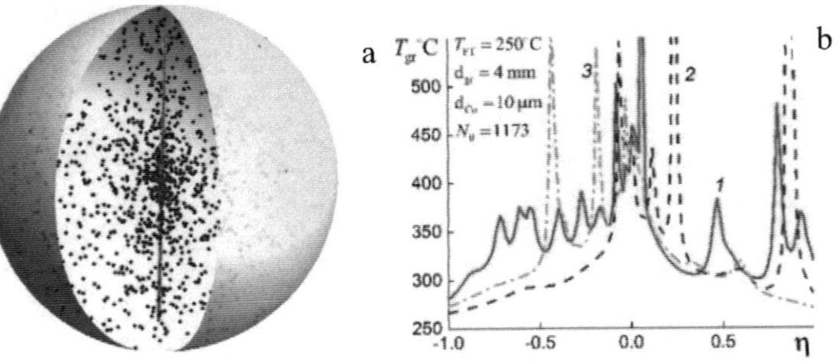

Figure 3. Distribution of active microparticles with a maximum concentration in the central part of the granule (a) and temperature distribution along three randomly selected diameters (b).

Using different densities of microparticles inside the granule, it is possible to control the temperature distribution in the granule volume. Figure 3 shows that an increase in the concentration of active microparticles in the center of the granule leads to a noticeable increase in the temperature in the center compared to the periphery of the granule.

The loss of thermal stability of the granule as a whole is determined by the diameter of the granule and the temperature of the reactor. Increasing the diameter of the granule reduces the critical temperature of the reactor, after which a thermal explosion occurs. With a fixed diameter of the catalyst granules, there is a critical temperature of the reactor, exceeding which will always lead to a thermal explosion.

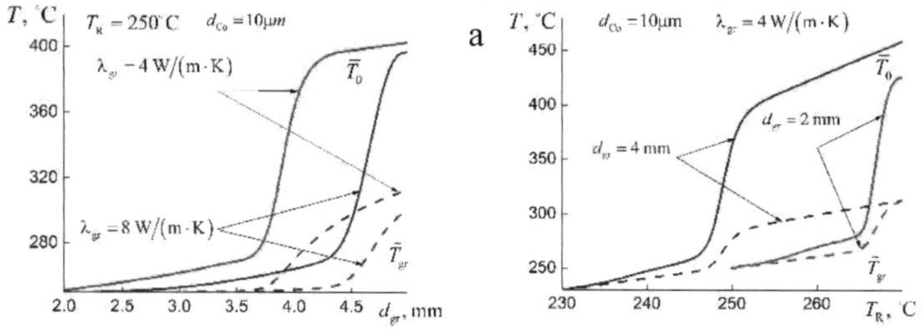

Figure 4. Loss of thermal stability of the granule as a whole, depending on the diameter of the granule (a) and the temperature of the reactor (b).

Figure 4 shows the dependences of the average temperature of the granule and the average temperature of the active centers with the loss of thermal stability of the granule. It can be seen that the temperature of the active centers is always higher than the surface temperature of the granule.

5. Conclusions

The method for calculating the heat and mass transfer in porous granules of a synthesis catalyst with point reaction centers has been proposed. The rate of an exothermic chemical reaction has been described by a modified Arrhenius law. The unknown temperature and concentration of the reagent at the active reaction centers has been calculated in a self-consistent iterative way. For the catalyst granule, we distinguish two types of thermal explosion. During a microexplosion, the active point centers of the synthesis reaction are significantly overheated. At the same time, the temperature of the granule volume and its surface is significantly lower than the temperature of the active centers. The thermal explosion of the catalyst granule as a whole is accompanied by a sharp increase in the average temperature of the granule. At a given reactor temperature, there is a critical granule size, the excess of which entails a thermal explosion. An increase in the reactor temperature also contributes to the onset of a thermal explosion.

Acknowledgments
This work was supported by the grants of Russian Foundation for Basic Research No 20-08-01061.

References

[1] Ahn C I, Bae J W 2016 *Catal. Today* **265** 27
[2] Wei L, Zhao Y, Zhang Y, Liu C, Hong J, Xiong H, and Li J 2016 *J. Catal.* **340** 205
[3] Khodakov A Y, Griboval-Constant A, Bechara R and Zholobenko V L 2002 *J. Catal.* **206** 230
[4] Ohtsuka Y, Arai T, Takasaki S and Tsubouchi N 2003 *Energy & Fuels* **17** 804
[5] Bhatia S K, Perlmutter D D 1983 *A.I.Ch.E. J.* **29** 281
[6] Benedetti A, Strumendo M 2015 *Chem. Eng. Trans.* **43** 1153
[7] Morse P M, Feshbach H, 1953 *Methods of theoretical physics. Part I* (New York: McGraw-Hill)
[8] Samarskii A A, Nikolaev E S 1978 *Metods of solving the grid equations* (Moscow: Nauka)
[9] Ermolaev V S, Gryaznov K O, Mitberg E B, Mordkovich V Z and Tretyakov V F 2015 *Chem. Eng. Sci.* **138** 1
[10] Wang Y, Jiang Y, Huang J, Liang J, Wang H, Li Z, Wu J, Li M, Zhao Y and Niu J 2016 *Fuel* **174** 17

Investigating the characteristics of the development of the boundary layer motion of locally nonequilibrium systems

V I Popov and A V Kuznetsov

Kutateladze Institute of Thermophysics SB RAS, Novosibirsk, Russia

vi1popov@mail.ru

Abstract. The method of integral momentum transfer relations has been extended to polymer systems with a locally nonequilibrium relaxation microstructure. The influence of the locally non-equilibrium transfer of the impulse flux on the characteristics of the development of their boundary-layer motion is analyzed.

Introduction

The method of integral relations is often used in the analysis of the integral characteristics of boundary layer motion, resistance and heat and mass transfer of ordinary structureless media [1,2]. This is due to the mathematical difficulties that arise in the joint integration of nonlinear differential equations of boundary layer motion and heat and mass transfer, despite the simplest (instantaneous and local equilibrium) connections that close the dynamic equation with respect to the flow and the associated forces.

Locally nonequilibrium media and their systems are endowed with a complex internal polymer microstructure and, as a result, have special physical and chemical properties [3-5]. Depending on the deformation conditions (flow geometry), they exhibit nonlinear-viscous properties, partially store the energy, supplied from outside in the form of changes in the components of the normal stress tensor, and relax the stress state, exhibiting non-instantaneous and locally nonequilibrium [6].

Applied problems on the influence of locally nonequilibrium transfer processes on the boundary layer movement and heat and mass transfer arise in the technological processes of polymer coatings of products from corrosion by immersion, in flows, when protecting the surface from icing, in the initial sections in the channels of extrusion molding, as well as when obtaining and modifying stratified flow solutions and melts of polymers and their systems.

The importance of the problem is complemented by the fact that when obtaining a high-quality adhesive coating and a uniform distribution of modifying micro and nano-additives with a predicted microstructure, it is necessary to have parameters for matching the technological process time and the relaxation time of polymer systems.

In contrast to structureless media, the problem due to relaxation phenomena in polymer systems is additionally complicated by the fact that the systems under consideration respond nonlinearly to external disturbances. In addition, due to delayed relaxation phenomena, they respond immediately in time and nonlocally, that is, nonequilibrium even in a small differential volume (point) of the medium, the deformation state of which depends on the state of the nearest medium[6]. This difference arises at the level of rheological closure of the law of conservation of momentum (momentum) of the elementary volume of the system.

Content from this work may be used under the terms of the Creative Commons Attribution 3.0 licence. Any further distribution of this work must maintain attribution to the author(s) and the title of the work, journal citation and DOI.
Published under licence by IOP Publishing Ltd

The regularities of motion in the case of an external flow of non-equilibrium media with a complex internal structure around bodies have been insufficiently studied. The problem lies, as noted, in the closure of the fundamental conservation laws by the equation of state of the medium, in the development of engineering methods for calculating the transfer of momentum, which allows analyzing the influence of structural and relaxation properties on the transfer processes.

The main goal of the work is to develop an applied research method for the study of quasi-stationary boundary-layer processes of transfer of locally nonequilibrium polymer media and their systems and to establish patterns and analyze the influence of locally nonequilibrium processes of momentum transfer on the development and characteristics of their boundary layer motion.

2. Integral ratio of impulses of boundary layer motion of locally nonequilibrium systems.

The starting point of this work is the analysis of the fundamental system, which is not closed with respect to flows and coupled forces, and the dynamic equations for the stationary boundary layer motion of locally nonequilibrium polymer media and their systems.

In the boundary layer theory,the well-known estimates [1,2,7] of the terms included in the dynamic equations with respect to the layer length and thickness of development lead to a system of equations of two-dimensional stationary boundary layer motion in the form:

$$\rho(u\partial u/\partial x + v\partial u/\partial y) = \partial P/\partial x + \partial P_{xx}/\partial x + \partial P_{xy}/\partial y \quad (1)$$
$$\partial P/\partial y + \partial P_{yy}/\partial y = 0$$
$$\partial u/\partial x + \partial v/\partial y = 0$$

The flow field in the outer region of the streamlined body with velocity u0, as is usually assumed in the theory of boundary layer motion, is described by the solution of the Euler equation.

Differentiating the second equation of system (1) with respect to x, from the first two equations we have:

$$\rho(u\partial u/\partial x + v\partial u/\partial y) = \partial P_{xy}/\partial y + \partial/\partial x\ (P_{xx} - P_{yy}) \quad (2)$$

Equation (2) was obtained taking into account the concepts of pressure continuity and constancy of velocity of the gradientless boundary layer motion, zero value of the velocity gradient, shear stress gradient, and gradient of the difference between the normal components of the deviator of the stress tensorin the outer region.

Equation (2) is not closed with respect to the kinematic characteristics of the flow and requires the introduction of the rheological equation of state. However, already in this form it establishes that the forces of inertia in the boundary layer are balanced by the gradients of shear stresses and the first difference of the normal components of the deviator of the stress tensor [7,8].

It should be noted that, in contrast to ordinary structureless fluids, for locally nonequilibrium systems, the value of P in (1) is not a hydrostatic pressure. This is an arbitrary scalar quantity that cannot be regarded as an isotropic pressure, since the trace of the stress tensor for the media under consideration and their systems is not equal to zero. In equation (2), this uncertainty is eliminated by the manifestation of the difference P_{xx} -P_{yy} of the normal components of the stress tensor.

Equation (2) is the starting point for deriving the integral boundary layer ratio of impulses. Omitting the characteristic calculations known for ordinary media [1,2,9], we obtain an equation for a gradientless flow of pulses in the form:

$$\rho u_0^2(d\delta^{**}/dx) = P_{CT}(x) - d/dx \int_0^{\delta}(P_{xx} - P_{yy})dy \quad (3)$$

Equation (3) containsthe pulse loss thickness $\delta^{**} = \delta\int_0^1 \omega(1-\omega)d\xi$ is pulse loss thickness, and $\xi = y/\delta(x)$which is the dimensionless coordinate, divided by the thickness of the boundary layer $\delta(x)$. The quantity $\omega = u/u_0$, where u0 is the velocity of the flow unperturbed by the wall; ρ is the density of the medium; $P_{CT}(x)$ is the shear stress on the wall (y=0) variable along the longitudinal coordinate x, and P_{xx}-P_{yy} is the first difference of the normal components of the stress tensor deviator.

Thus, the problem is reduced to closing the ratio of the momentum flux, establishing the connection between the components of the stress tensor and the components of the strain rate tensor, which are the main kinematic characteristics of the flow. That is, the problem is reduced to the establishment of the rheological equation of state between the flows and the coupled forces in relation to the integral ratio of momenta for polymer media and their systems.

3. The rheological equation of state for the boundary layer transfer of the impulse flux of locally nonequilibrium systems.

At present, phenomenological and structural-statistical methods and approaches are being developed for the closure of fundamental conservation laws. In [10,11], within the framework of extended nonequilibrium statistical thermodynamics for a class of polymer media, a non-instantaneous and non-local process of transfer of a momentum flux of polymer media with a relaxation microstructure in the field of shear, entropy, and diffusion forces was modeled. The relations for the dynamic components of the stress tensor of the pulse flow with the coupled forces are obtained in the form of:

$$P_{xy} = \frac{\varepsilon G æ}{2}(1 - \exp(-2æ^{-1}t)) \quad (4)$$

$$P_{xx} - P_{yy} = \frac{\varepsilon G^2 æ^2}{2}[1 - \exp(-2æ^{-1}t) - 2æ^{-1}t\exp(-2æ^{-1}t)] \quad (5)$$

The analysis of the space-time relations of equations 4 and 5 (Figs. 1 and 2) reveals the areas of influence of the shear and the first difference of the normal components of the stress tensor depending on the properties of the medium and the conditions of uniform deformation $G \equiv du/dy$. The criterion for the non-instantaneous influence is the Deborah number (the ratio of the relaxation time æ to the characteristic time of the process $De = æ/t$), and locally, the non-equilibrium influence is the Weissenberg number (We = Gæ). The value $0.5\varepsilon æ = \mu$is the dynamic viscosity, and the quantity ε is the elastic modulus. Equations (4 and 5) also lead to the fundamental conclusion that the variable non-Newtonian viscosity of the medium (dependence on the shear rate) is due to relaxation phenomena, which determines the processes of momentum transfer aimed at neutralizing external disturbances.

The areas of influence of local nonequilibrium are establishedin accordance with (5),(Fig. 2). At We> 2, dissipative rheological flows are not determined only by the gradients of the corresponding transfer potential, as for ordinary structureless media, but are already the solutions of an evolutionary equation describing the process of relaxation of the stress state parameters of a nonequilibrium polymer system to their local equilibrium values [11].

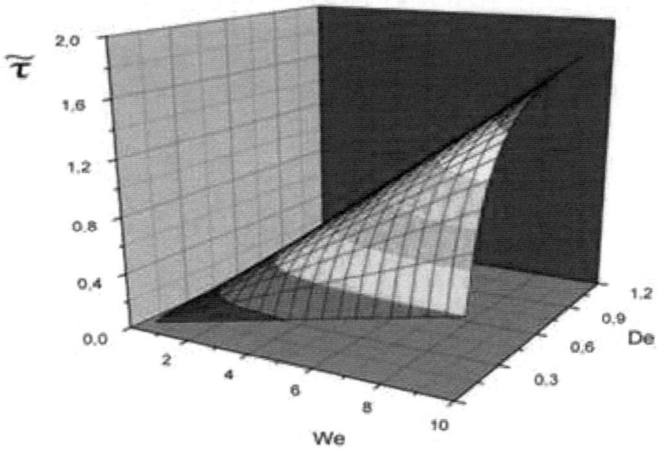

Figure1. Influence of De and We on the magnitude and nature of the change in the shear component of the stress tensor in a locally nonequilibrium process of momentum transfer.

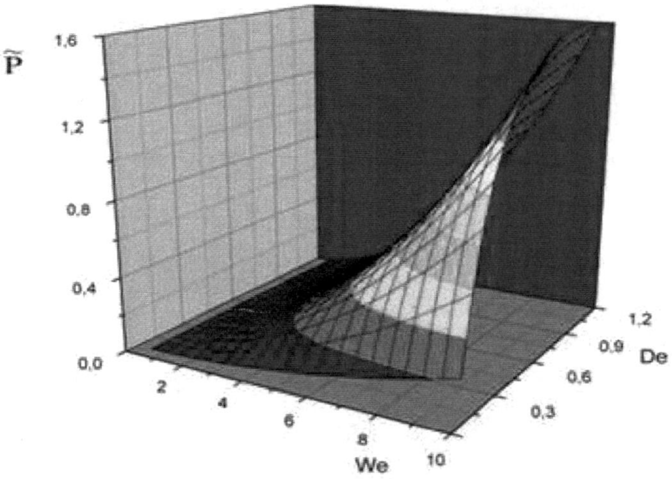

Figure 2. Influence of De and We on the magnitude and nature of changes in the first difference of the normal components of the stress tensor in a locally nonequilibrium process of momentum transfer.

It follows from relations (4 and 5) that under the conditions of a quasi-stationary shear characteristic of the integral boundary layer class of problems (3), the stressed state of a medium with a locally nonequilibrium relaxation microstructure has the form:

$$\frac{P_{xx}-P_{yy}}{P_{xy}} = G æ \qquad (6)$$

4. Influence of the locally nonequilibrium process of momentum transfer on the characteristics of the development of boundary layer motion.

In the method of integral relations, it is usually assumed that the velocity field is approximated across the boundary layer, followed by the determination of its thickness and the corresponding characteristics of the flow along the length of its development. The distribution of velocities is given from the boundary conditions of the flow [1,2,9]. In this case, an assumption is made about the affinity of the velocity profiles on a flat plate streamlined in the longitudinal direction.

This workinitiallyusedthe distribution of not the kinematic, but the dynamic characteristics of the flow over the cross section of the boundary layer by a cubic parabola [12]. The advantage of this approximate approach is that it does not depend on the rheological properties of the medium under study. The values of the coefficients are determined from the boundary conditions of the problem $\xi = 0, \tilde{P}_{xy} = 1; \xi = 1, \tilde{P}_{xy} = 0$.

$$\tilde{P}_{xy} = \frac{P_{xy}(x)}{P_{CT}(x)} = 1 - 3\xi^2 + 2\xi^3 \qquad (7)$$

The distribution of the longitudinal component of the velocity vector is determined from the quasi-stationary ($t \to \infty$) relation of the system of equations (4) and (7):

$$\omega = \frac{u}{u_0} = \frac{2P_{CT}(x)\delta}{\varepsilon æ} u_0 \int_0^{\xi} \tilde{P}_{xy} d\xi \qquad (8)$$

Therefore, from the boundary conditions ξ=1, ω=1, we find the relation for $P_{CT}(x)$ and the distribution for the velocity profile across the boundary layer:

$$\omega = 2\xi - 2\xi^3 + \xi^4 \qquad (9)$$

Substituting the corresponding values of quantities in relation (3), at $\delta^{**} = 0{,}1175\delta$, awe arrive at a closed momentum equation for changing the thickness of the boundary layer along the longitudinal coordinate x:

$$\frac{d\delta}{dx} + \frac{12.65 \nu æ d\delta^{-1}}{dx} - \frac{17.02}{\delta u_0} = 0 \qquad (10)$$

where $v = \mu/\rho$ is the kinematic medium viscosity.

The solution to equation (10), considering x = 1,δ = 0, has the form:

$$\delta^2 - 25.3v\text{æ}\ln(\delta) - \frac{34.04v}{u_0(x+1)} = 0 \quad (11)$$

From equation (11) it follows that there is a class of locally nonequilibrium media for which the inertial forces of equation (2) are balanced simultaneously by tangential and normal stresses (Fig. 2, the number We> 5), and the boundary layer does not develop from the leading critical edge of the flow, but at some distance from it.

The development of the layer from the leading-edge x = 0 of the flow decelerationcan occur after the completion of dissipative (G P_{xy}) relaxation from a quasi-solid to some fluid state of stresses in the flow (the number We <5). In this case, the last term in equation (2) is no longer dominant, and the inertial forces are balanced by the shear stresses and partly by the first difference of normal stresses.

It should be noted that this relaxation process of unwanted elastic turbulence into the pre-inlet region of the forming channel [13] can be observed in the pre-inlet chambers of the channels of the forming elements in the extrusion technology, the source of which is their leading edges of the flow deceleration.

In accordance with relation (6) and constancy of We, to study the influence of the locally nonequilibrium process of momentum transfer on the characteristics of the boundary layer motion, starting from the leading edge of the plate, it is convenient to use the cubic parabola approximation (7).

For a locally nonequilibrium flow, when the last term in equation (2) is not dominant (2 <We <5) from relation (3) we have:

$$\delta^2 + \frac{17.02\delta We}{Re} - \frac{34.04X}{Re} = 0 \quad (12)$$

From equation (12) we find the development of the boundary layerthicknessalong the plate of length L, that is (X = x / L), depending on the numbers$Re = u_0 L/v$ and We. The criterion We / Re, which depends only on the rheological properties of the medium and, essentially, on the geometry of the flow, for the first time offered in [8] based on the analysis of dimensions,is proposed for systematic use. In equation (12),δ is dimensioned by the length L.

The change in the thickness of the boundary layer along the flow, considering δ = 0 at X = 0, has the form:

$$\delta = 8.51We/Re[\left(1 + \frac{0.47XRe}{We^2}\right)^{0.5} - 1] \quad (13)$$

For the case of a locally equilibrium flow (We/Re → 0), from Eq. (12) we have changes of the thickness of the boundary layer, shear stress and local resistance in the form [12]:

$$\delta(x) = 5.83\left(\frac{vx}{u_0}\right)^{0.5} \quad (14)$$

$$P_{CT}(X) = \frac{\mu u_0}{\delta(x)\left(\frac{d\omega}{d\xi}\right)_{CT}} = 0.343\left(\frac{\mu\rho u_0^3}{x}\right)^{0.5}$$

Local resistance coefficient in section x:

$$C_x = \frac{2P_{CT}(X)}{\rho u_0^2} = 0.686(v/xu_0)^{0.5} \quad (15)$$

5. Conclusion

The method of integral relations has been extended to the class of polymer media and their systems with a locally nonequilibrium relaxation microstructure. The problem of stationary flow around a plate has been considered. The characteristics of nonequilibrium influence on the development of the boundary layer have been established and analyzedfor the identified areas of locally nonequilibrium deformation. The development of the boundary layer depends on the properties and essentially on the geometry of the flow. In locally nonequilibrium systems with a relaxation microstructure, the transfer processes are characterized not only by boundary potentials, but also by relaxation phenomena in subsystems with a delay to the equilibrium state, affecting the development of the boundary layer.

References

[1]. Schlichting H. Theory of the boundary layer. M.: Science, publishing house Phys.- mat., 1969.
[2]. Loytsyansky L.G. Laminar boundary layer. M.: publishing house Phys.- mat., 1962.
[3]. Bird R.B., Armstrong R.C., Hassager O. Dynamics of polymeric liquids: Fluid Mechanics. New York:Wiley - Interscience, 1987.
[4]. Vinogradov V.G., Malkin A. Ya. Polymer rheology. Moscow: Chemistry, 1977.
[5]. Middleman S. Polymer flow. Moscow: Mir, 1971.
[6]. Popov V.I. The role of nonlocality and delay effects in transport processes in media with a microstructure. Journal of PMTF. 2002. V. 43. No. 6, p. 151.
[7]. Metzner A.B. WhiteJ.L. Flow Behavior of Viscoelastic Fluids in the Inlet Region of a Channel. AIChE Journal, 1965. V.11, No. 6. p. 989.
[8]. Astarita J., Marruchi J. Fundamentals of hydromechanics of non-Newtonian fluids. M: Mir, 1978.
[9]. Targ S.M. The main tasks of the theory of laminar flows. M.: Publishing house of Techniko - theoretical literature, 1951.
[10]. Popov V.I. Rheokinetics of substance and momentum transfer in chemically active media with a microstructure. Thermal physics and aeromechanics. 2001. Vol. 8, No. 2, p. 269.
[11]. Popov V.I. Evolution of chemically active polymer structures in the field of shear, entropy and diffusion forces. Journal of Theorist. fundamentals of chem. technologies. 2011. V.45. No. 5, p.519.
[12]. Kutateladze S.S. Foundations of the theory of heat transfer. Publishing house «Science». Siberian branch Novosibirsk, 1970.
[13]. Popov V.I. The emergence and development of elastic turbulence in polymer flows. Journal of Theorist. fundamentals of chem. technologies. 2019. V.53, No. 2, p. 189.

Application of machine learning to predict the thermal power plant process condition

M M Sultanov, I A Boldyrev and K V Evseev

Volzhsky Branch of Moscow Power Engineering Institute, ul. Lenina 69, Volzhsky, Russia

kirillevseyev@gmail.com

Abstract. This paper deals with the development of an algorithm for predicting thermal power plant process variables. The input data are described, and the data cleaning algorithm is presented along with the Python frameworks used. The employed machine learning model is discussed, and the results are presented.

1. Introduction

Building a digital system for predicting process conditions leading to thermal power equipment malfunctions is crucial for providing high operation efficiency and safety of thermal power plants. Malfunctions can be both direct technological (vibrations, turbine metal temperature, etc.) and general (equipment condition, fuel-consumption rate etc.). To predict such parameters, it is proposed to use time series forecasting techniques since the data in this case are an array of technological values together with timestamps [1–3].

The goal of this study is to develop an online algorithm for forecasting thermal power plant process variables to improve the plant efficiency and reliability.

The input data were retrieved from a historical database and contain process variables of a power plant turbine set, such as the sharp steam flow, the temperature and pressure behind the boiler, and the temperature of the various metal parts of the turbine set. The raw data contain approximately 41000 entries over 25 days.

The algorithm was developed using Python, which is a high-level programming language that provides a flexible environment for data analysis and machine learning. The following major Python frameworks and libraries were used to process the data:

— Numpy (for working with array data);
— Pandas (for working with panel and time-series data);
— Scikit-learn (for preprocessing data to fit a machine learning model);
— Matplotlib (for data visualization and presentation);
— Keras (for building and fitting machine learning models).

2. Data preparation

The raw input data are saved as a comma-separated values (CSV) file, where each column represents a parameter value and the index is in date/time format. The column names correspond to the parameter name signal markings, which, though less conveniently representable, are easier to access when

Content from this work may be used under the terms of the Creative Commons Attribution 3.0 licence. Any further distribution of this work must maintain attribution to the author(s) and the title of the work, journal citation and DOI.
Published under licence by IOP Publishing Ltd

addressing specific values in the data table. Signals are represented using a dictionary which contains markings as keys and parameter names as values, making it possible to get a user-friendly description of a particular signal marking when analyzing and visualizing data.

The first step in analyzing process variables is to join all parameters by timestamps using outer merge. Since the data is stored in the software-and-hardware historical database at different time intervals, i.e., some parameters can be written more often, the next step is to fill the empty fields with the last known states, assuming that the parameter value has not been updated yet [4]. However, after the join, the time interval between the entries may be different. Therefore, to use the timeseries in forecasting, it is necessary to time sample the data at 1-min intervals, so that the timestamp delta for each field is the same and equals one minute. An example of a formatted data table is presented in Figure 1.

date	3LBA10CF021	3LBA20CF022	3LBA30CF023	3LBA40CF024	3LBA20CP003	3LBA30CP001A	3LBA70CP001A	3MAA10CT001
2009-01-12 22:00:00	50.4	34.8	7.3	39.9	27.7	0.8	0.7	80.6
2009-01-12 22:01:00	57.2	34.8	7.3	39.9	27.7	0.8	0.7	81.4
2009-01-12 22:02:00	65.2	34.8	7.3	39.9	27.7	0.8	0.7	81.4
2009-01-12 22:03:00	63.0	34.8	7.3	39.9	27.7	0.8	0.7	81.4
2009-01-12 22:04:00	60.5	34.8	7.3	39.9	27.7	0.8	0.7	81.4

Figure 1. Process variable data table

Each parameter can be represented as a plot where the X-axis is the timestamp and the Y-axis is the parameter values. The turbine set process variable corresponding to the metal temperature in the steam inlet zone in the lower half of the high-power cylinder (HPC) is shown in Figure 2.

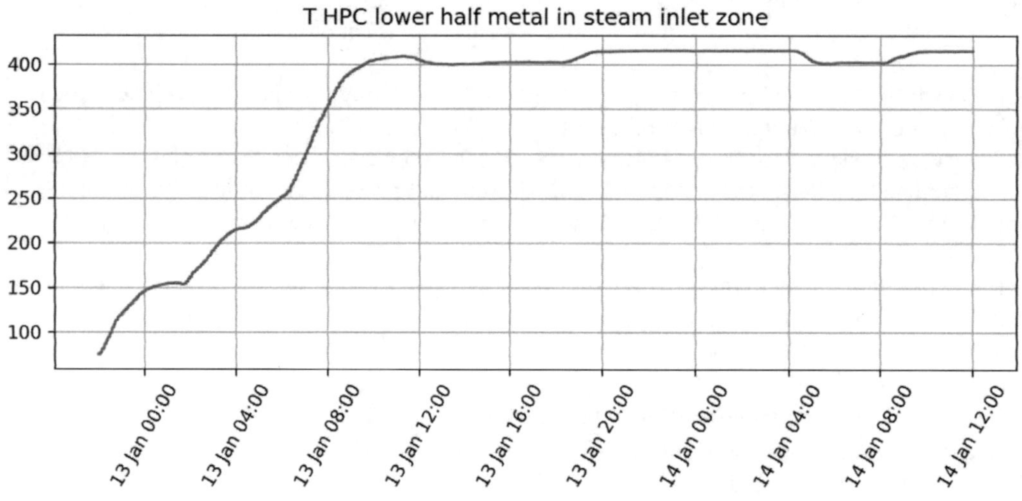

Figure 2. Process variable plot

The parameter presented above is one of the target parameters to be monitored during thermal power operation, especially when starting or stopping a turbine set. The plot represents a transient process, a case where forecasting process variables is important for providing reliable and efficient operation.

3. Process variable forecasting

Process variable forecasting is needed to prevent malfunctions and improve the power equipment efficiency. To make a forecast, historical data close to the period of prediction are required. Any forecasting algorithm should be validated before it can be used to predict future values, so that when forecasting a process variable, it is necessary to split the input data into training and validation sets. In fitting a model, the training set is used to find the optimal model weights and the validation set is used to estimate the error in predicting new values. The validation set error is calculated at the end of an epoch end and is a metric that indicates whether the model is overfitted and allows making a decision when to stop fitting the model. The study validation set is 10% of the dataset used to predict future values.

To predict the future values are predicted using a forecast window which contains a number of previous values. The size of the forecast window depends on the parameter. The prediction interval corresponds to the number of future values to be forecasted using the values within the forecast window. To predict the HPC metal temperature within 15 to 30 min, it is suggested to take into account the values for the last 30 min. to 1 hour. The forecast dataset should include several hours before the forecast point.

To forecast process variables, it is proposed to build a machine learning model consisting of two long short-term memory (LSTM) blocks (Figure 3). This structure is due to the fact that values of a process variable depend on its previous states. Moreover, it is proposed to consider not only the regressor, the process variable itself, but also include other variables, predictors that are shifted in time. Therefore, to predict future values, it is necessary to use the values of both the regressor and predictors within the forecast window. The proposed model takes into account the values for the last 60 min. and predicts the future values of the regressor for 30 min, i.e., its output contains 30 values.

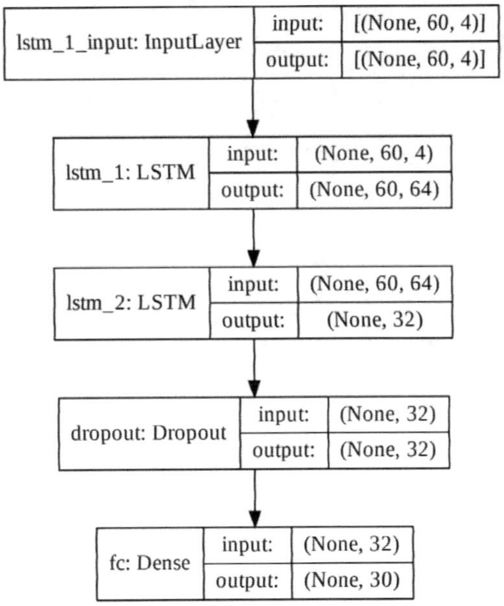

Figure 3. Forecast model structure

Before using data to fit the model, they should be normalized to avoid large and small errors and to speed up the training phase. To fit the model, both the regressor and predictors should be split into arrays in which each element is a window containing the corresponding values. The error between the ground true and predicted values is calculated using the mean absolute deviation (MAD). During model training, it is necessary to avoid over fitting. For this, at the end of each epoch, the validation loss is compared with the minimum loss obtained and the best model weights are stored in memory. After the model fitting, the best weights are restored to make predictions.

It should be noted that the model should be fitted each time before forecasting using the last data obtained. This is due to the fact that the data should be close to the forecast point to make correct predictions and new data has their own mean and standard deviations, so that reverse normalization may return a false result. To provide the algorithm performance the model should be fitted at least every 5 minutes. The model predictions are used to evaluate the process and take actions to improve the unit performance. The loss metric graph during the training phase is presented in figure 4, where each epoch corresponds mean squared error for both training and validation data sets.

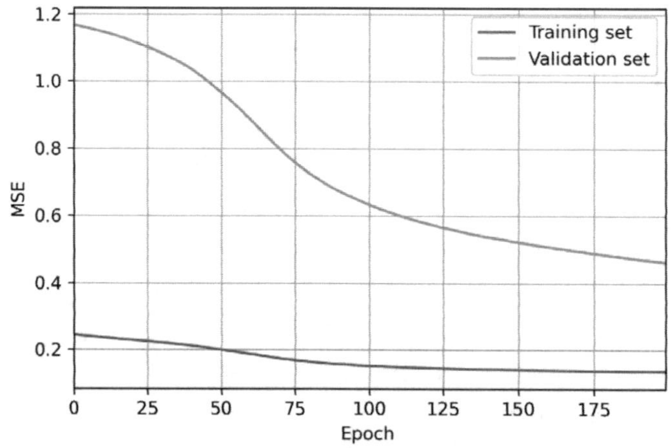

Figure 4. Training loss graph

The results for the metal temperature in the steam inlet zone in the HPC lower half are presented in Figure 4. The dashed line corresponds to the ground truth, and the solid line to predictions. The vertical line at 6:00 separates the training data on the left from the prediction interval on the right. To plot the predictions for training dataset the first values of the array of predictions that returns the model were taken. The reason to plot these predictions is model performance visualization over both training and testing datasets.

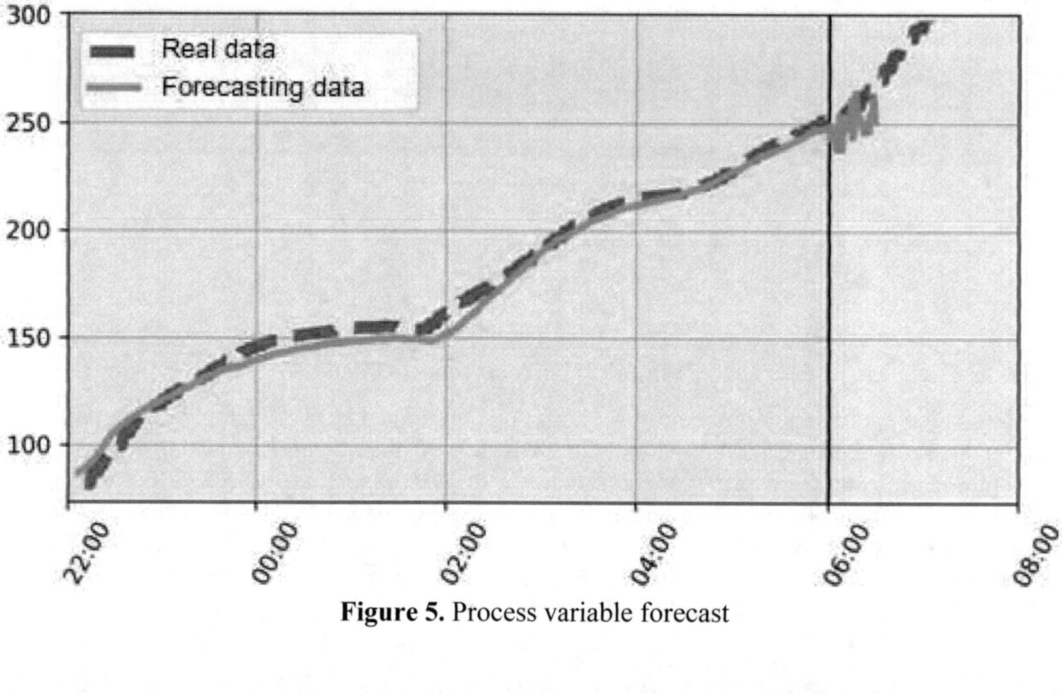

Figure 5. Process variable forecast

4. Conclusions

The proposed algorithm allows forecasting process variables taking into account the previous state of both the regressor itself and predictors; therefore, the prediction is based on multivariate time series forecasting. The forecasting pipeline provides automation of the future value prediction, which allows one to analyze data online, retrieve the last acquired values, and return predictions within a certain time interval as feedback. The use of the proposed solution allows improving the reliability and efficiency of thermal power plants by preventing inefficient generating unit operation.

Acknowledgements

The research is funded by Russian Federation public contract FSWF -2020-0025 «Technique development and method analysis for ensuring power system object security and competitiveness based on the digital technologies».

References

[1] John Harlim, Shixiao W. Jiang, Senwei Liang, Haizhao Yang. Machine learning for prediction with missing dynamics, Journal of Computational Physics, 2021, Vol. 428, 109922

[2] Wang, Z. Zhao, B. Guo, H. Tang, L. Peng, Y. Deep Ensemble Learning Model for Short-Term Load Forecasting within Active Learning Framework, Energies, 2019, Vol. 12(20), 3809.

[3] Arakelyan, E.K., Sultanov, M.M., Boldyrev, I.A., Gorban, Y.A., Evseev, K.V. Analysis of the DCS historical data for estimation of input signal significance, Proceedings of the 3rd 2021 International Youth Conference on Radio Electronics, Electrical and Power Engineering, REEPE 2021, 2021, 9388069

[4] Arakelyan E.K. Application of machine learning methods for optimizing technical and economic performance of generating systems / Arakelyan E.K., Boldyrev I.A., Evseev K.V., Gorban Yu. A. // IOP Conference Series: Materials Science and Engineering, №1035, 2021

Experimental studies of the formation of zones resulted from air flow around a system of model buildings

V D Meshkova[1], A A Dekterev[1,2], D A Dekterev[1,2] and A S Lobasov[1,2]

[1] Siberian Federal University, Krasnoyarsk, 79 Svobodny Avenue, 660041, Russia
[2] Kutateladze Institute of Thermophysics, Siberian Branch, Russian Academy of Sciences,. 1 Lavrentyev Avenue, Novosibirsk, 630090 Russia

redel-vd@yandex.ru

Abstract. This paper presents the results of working out the methodology for conducting experimental studies of the flow around the objects modeling the urban environmental conditions. The experiments were conducted in the wind tunnel of the Siberian Federal University. Two objects of different heights imitating buildings were considered the models. Special attention was paid to the study of the flow pattern at the tandem arrangement of model buildings. Visualizing the flow, the low-velocity and high-velocity zones, as well as recirculation areas were identified. At that, these zones had their peculiarities in terms of the direction of flow twisting behind each object. The study allowed revealing that the vortices separating from the edges of the studied objects play a special role in the flow formation.

1. Introduction

The main task of urban planning is not only to provide the population with living space but also to ensure that it meets the requirements of general and environmental safety and is comfortable from the perspective of citizens staying in it.

As it has been repeatedly noted, the contemporary location of buildings in the urban neighborhood units leads to a significant decrease in the aeration regime of the territory [1,2]. Dense development, i.e. the close location of buildings to each other contributes to the formation of many local zones with low wind velocities, which results in decreasing the dispersion of pollutants in the atmosphere, leading to the opposite effect, that is, their accumulation [3].

Contemporary approaches in architecture and urban planning allow the construction of buildings of various geometric complexity and height. The peculiarity of this choice leads to the fact that high-rise buildings form flow acceleration zones around themselves, which lead not only to the undesirable transfer of pollutants to high altitudes but also are very uncomfortable for people [4].

There are many studies of the wind flow effects on individual model structures, but most of these studies are of private concernment. Currently, no specialized experimental base is available for conducting physical modeling of the flow around urban planning elements and evaluating them from the standpoint of environmental safety and comfort.

The main problem is that a group of buildings in real conditions interacts with a set of vortex structures that arise as a result of the interaction of the air flow with the underlying surface and buildings, located upstream. Also, the problem of temperature influence on the flow remains unsolved, since when considering high-rise buildings, not only changes in wind velocity and direction are observed, but also

Content from this work may be used under the terms of the Creative Commons Attribution 3.0 licence. Any further distribution of this work must maintain attribution to the author(s) and the title of the work, journal citation and DOI.
Published under licence by IOP Publishing Ltd

the temperature of the flow itself changes both from outside (the environment) and inside (the building). If studying not only the modes of isothermal flow around buildings but also evaluate the effect of real heat sources, the analysis becomes even more complicated due to the significant mutual influence of aerodynamics and heat transfer.

To understand the current conditions, it is necessary to identify the main patterns of interaction of the flow with the elements of the urban environment. As a rule, a numerical study allows demonstrating clearly the spatial structure of the flow for various configurations of districts and building structures, as well as quickly analyzing their changes. However, the main question concerns the degree of adequacy of the results obtained in numerical simulation, since for their validation it is necessary to have data from field and laboratory experimental measurements, which are extremely poorly presented in research articles.

To solve this problem, this publication presents an initial stage of work that will allow further approaching a comprehensive solution to the problems under consideration.

The purpose of the scientific research was to develop methods and approaches in carrying out experimental work by determining the characteristic zones formed as a result of air flow around two closely located models of buildings that simulate real structures.

2. Problem statement

A physical experiment was conducted to work up experimental techniques and verify the numerical model. The concept of selecting research objects was based on the study [5].

Two geometric models of different heights, externally imitating buildings with a flat roof, were considered as the research objects (Fig. 1):

- cube of 50x50x50 mm (low-rise building);
- parallelepiped of 50x50x100 mm (multi-rise building).

The option of their tandem arrangement was considered. The models were made of optically transparent Plexiglas. The facets located on the observation side of the camera were covered with an opaque film to reduce the glare sources (fig. 2).

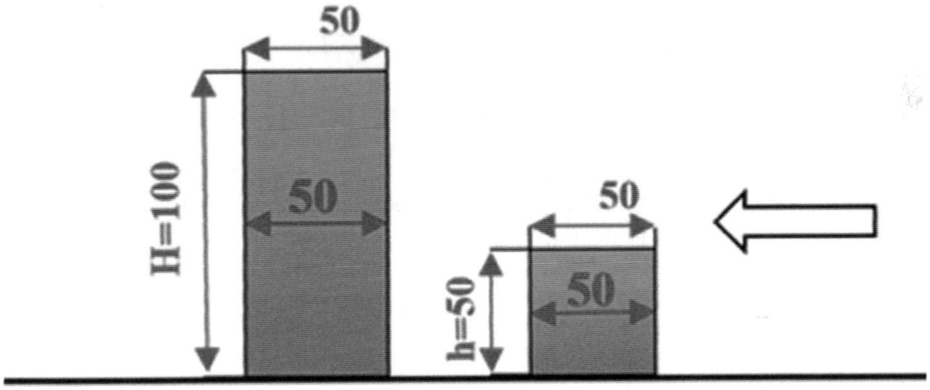

Figure 1. The characteristic dimensions of the objects under study.

Figure 2. The studied objects.

The incoming flow velocity was 5 m/s (the corresponding Re=16,000 calculated by a characteristic size equal to the height of the smaller model of 50 mm).

3. Approaches and methods

The experimental study was carried out in the wind tunnel of the Siberian Federal University, which can be attributed to the class of low-velocity open-type wind tunnels with a closed working part. Structurally, the wind tunnel consists of the following elements: a fan, a diffuser, a working section, and a nozzle. Visualization during the flow around the studied objects was implemented using a continuous diode laser.

A micro-scale model of atmospheric air based on the Reynolds-averaged Navier-Stokes equations for incompressible flows with variable density was used for numerical simulations. The system of basic equations includes the equations of continuity, motion, and energy conservation. To close the equations in the turbulent regime, the k-ω SST model was used [6]. The simulation was carried out employing a SigmaFlow non-commercial universal software package developed by specialists of the Institute of Thermophysics SB RAS and the Department of Thermophysics of SFU [7]. The proposed numerical model was verified earlier using the data of experimental works of foreign scientists, and showed good consistency of the numerical simulation with the experimental results [8,9].

4. Analysis of the results obtained

The experimental results are presented in the form of flow visualization using talc. The choice of this substance is due to its physical properties, namely, the small size of 20-50 microns and low density of 2.58-2.83 g/cm^3 [10]. Talc particles do not cause additional disturbances into the flow and allow demonstrating the flow structure (Fig. 3a,b).

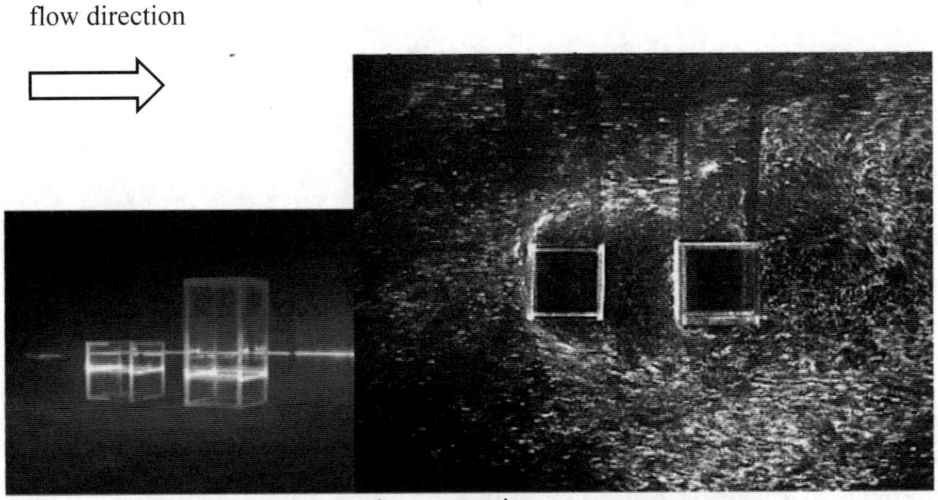

a) cross section

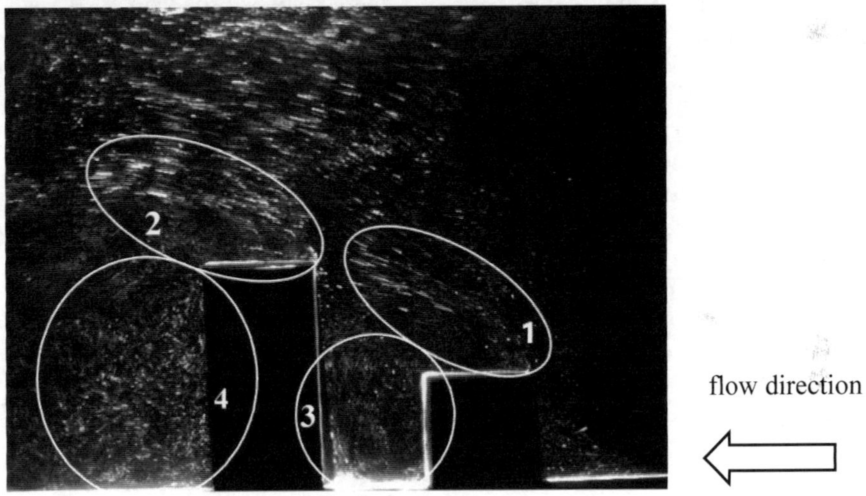

b) longitudinal section

Figure 3. Visualization of the flow structure when flowing around objects, using talc.

Visual observations of the flow around buildings have shown that complex non-stationary three-dimensional vortex flows and large-sized zones of separated flow with the formation of wave effects are formed around buildings, and especially from the windward side (zones 1,2). A stable recirculation area (zone 3) is formed between a low-rise and a high-rise building, where particles are practically not carried outside. An aerodynamic shadow is formed behind a high-rise building (zone 4), which also leads to the cycling movement (Fig. 3b).

The numerical study allows demonstrating that the qualitative, simulated, and experimental flow patterns have a similar character and reflect the appearance of all peculiar zones as a result of the interaction of the flow with the objects under study (Fig. 4).

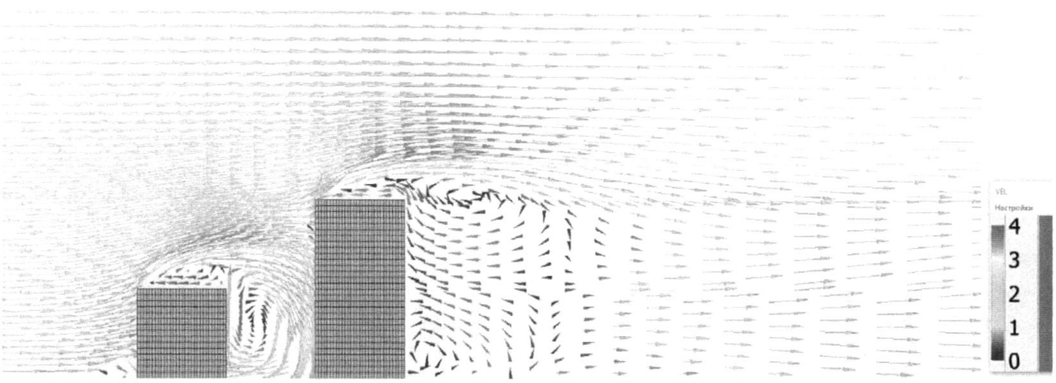

Figure 4. The results of numerical simulation of the flow around two objects, m/s

5. Discussion

The results of the work conducted allow demonstrating the formation mechanism of the characteristic zones when objects are streamlined by the wind flow, as well as assessing the degree of influence of both buildings on the change in the flow structure, and the influence of objects on each other. In the context of a dense location of the observed objects, a special role is played by the interaction of vortices separated from the edge of buildings, and the presence of separation and recirculation zones near buildings. At that, changes in the characteristic sizes of the objects under study and the distances between them lead to an increase or a decrease in the formation of a particular zone.

The present work is the first in the cycle of research, conducted in the field of computational modeling of the ecological situation in the urban environment and assessment of pedestrian comfort of citizens in the territories under consideration.

Acknowledgments

The study was carried out with the financial support of the Krasnoyarsk Regional Fund for the Support of Scientific and Scientific and Technical Activities within the framework of the Targeted Competition of Applied Scientific Research aimed at solving the problems of urban development under the project No. 2020080506564 "Research and assessment of the impact of vehicle emissions on the urban environment using computational and experimental modeling"

References

[1] Skobeleva H, Pilipenko O and Bulgakov A 2018 Methodology for assessing the comfort of an urban environment. 10.3311/CCC2018-119

[2] Stuck M, Vidal A and et al. 2021 Spectral-Element Simulation of the Turbulent Flow in an Urban Environment. Appl. Sci. **11**, 6472 (https://doi.org/10.3390/ app11146472)

[3] Yoshihide Tominaga, Ryuichiro Yoshie, and et al. 2005 Cross Comparisons of CFD Prediction for Wind Environment at Pedestrian level around buildings The sixth Asia Pacific Conference on Wild Engineering (APCWE-V) Seoul, Korea, 12–14

[4] Ali M, Al-Kodmany K 2012 Tall Buildings and Urban Habitat of the 21st Century: A Global Perspective
Buildings, **2**, 384-423 (doi:10.3390/buildings2040384)

[5] Soo-Jin Park and at all. 2020 Flow Characteristics Around Step-Up Street Canyons with Various Building Aspect Ratios. Boundary-Layer Meteorology**174** 411– 431

[6] Menter F R 1993 Zonal two equation k-omega turbulence models for aerodynamic flows AIAA Paper. **93** 2906, 21

[7] Dekterev A A et al 2003 Using SigmaFlow for the numerical study of technological objects // Computational technologies **8** 1 250-255.
[8] Meshkova V D, Filimonov S A and Shebelev A V 2018 Computational modeling of architectural and construction aerodynamics problems // IOP Conf. Series: Materials Science and Engineering **456**
(doi:10.1088/1757-899X/456/1/012079)
[9] Meshkova V D and et al 2020 SigmaFlow CFD code as a tool for predicting the wind environment around a group of buildings // IOP Conf. Series: Materials Science and Engineering **1675** 012119
(doi:10.1088/1742-6596/1675/1/012119)
[10] Overview of the talc market in the CIS. M .: InfoMine, 2010.

Features of the educational program «Nanotechnologies and nanomaterials in energy»: machine learning

D. D. Babenko, A. S. Dmitriev, P.G. Makarov, I.A. Mikhailova

National Research University «MPEI», Russia, 111250 Moscow, Krasnokazarmennaya, 14

asdmitriev@mail.ru

Abstract. The results of the development of new educational programs in the field of nanotechnology and nanomaterials in the energy sector, which have been developed and are actively used by the National Research University "MPEI", are presented. Modern nanomaterials and nanotechnological processes in traditional and alternative ("green") energy require new approaches, including statistical methods for the analysis and synthesis of experimental data and design options. For this reason, without the active use of machine learning methods, it is impossible to train qualified specialists in the field of promising energy problems and their solutions. Through teaching, research, and innovation, «MPEI» exceptional community pursues its mission of service to the nation and the world.

1. Introduction

The training of specialists in the direction of bachelor's and master's in the direction of «Nanotechnology and nanomaterials in energy» began at the NRU «MPEI», starting in 2005. Specialized work programs were developed, lecture courses were prepared, the Nanocenter «MPEI» was created, laboratory works were prepared, topics were developed and assignments for practical and seminars. At the initial stage of training bachelors and masters, in addition to the teachers of the MPEI, the leading specialists in Russia from MEPhI, MIPT, Moscow State University were invited to give lectures and conduct classes. Among the developed and prepared courses in the direction of training «Nanotechnology and nanomaterials in energy»: «Solid state physics (for nanotechnologists)», «Electrodynamics and nanophotonics», «Quantum and optical electronics», «Introduction to nanoenergy», «Functional nanosurfaces for energy and optoelectronics», «Elements and devices of nanotechnology», «Sensors, MEMS, NEMS and biosensors», «Functional energetic materials», «Surface investigation in vacuum and low temperatures», «Methods for diagnostics of nanosystems and nanomaterials», «Thermal processes in nanostructures» and others (Figure 1).

The established laboratories in the field of nanotechnology and nanomaterials make it possible not only to obtain new materials based on micro- and nanoparticles, carbon nanotubes and graphene, but also to create new hybrid nanofluids and nanocoimposites based on them. When preparing students and postgraduates in the direction of nanomaterials and nanotechnology, the following are used - a laboratory complex DSA25 Drop Shape Analyzer measures the static or dynamic contact angle with high precision and can also be used to determine the surface free energy by carrying out measurements with several liquids. The instrument provides rapid results for the wettability of a surface, for example for checking the quality of cleaned, pretreated, or coated sample; scanning electron microscope EM-

Content from this work may be used under the terms of the Creative Commons Attribution 3.0 licence. Any further distribution of this work must maintain attribution to the author(s) and the title of the work, journal citation and DOI.
Published under licence by IOP Publishing Ltd

Journal of Physics: Conference Series **2150** (2022) 012031 doi:10.1088/1742-6596/2150/1/012031

30AX PLUS, a planarGROW-2S carbon structure synthesis unit, an MS-200 mass spectrometer, an LFA 457 MicroFlash laser flash unit from Netzsch (Germany) is a modern high-precision instrument for measuring the thermal diffusivity and thermal conductivity of a wide range of materials, a viscometer (vibro viscometer) A&D SV-10, and other equipment (Figure 2).

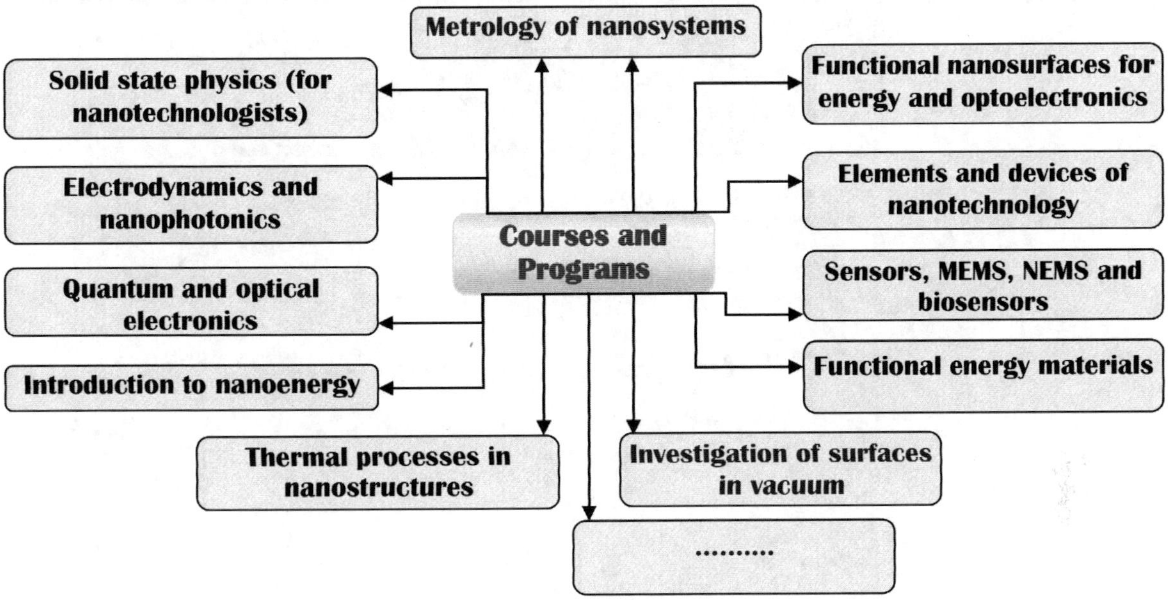

Figure 1. Courses and Programs

As part of the implementation of training programs, a significant number of studies were carried out on the functional surfaces of nanomaterials (wetting and spreading of droplets of working fluids, heat transfer coefficients and coefficient of resistance), as well as optical and absorption properties of surfaces (ceramics, nanocarbon composites, hybrid regular and random graphene nanocomposites and films, nanocolloids, etc.). Particular attention in the works of students and postgraduates is paid to such areas of research on nanomaterials and nanofluids as solar thermal energy, photovoltaic cells, lithium-ion batteries and supercapacitors, hydrogen fuel cells, thermal interface materials and heat storage systems [1-9].

Figure 2. Laboratory equipment

Partially the results of research within the framework of the program are presented in Figure 3.

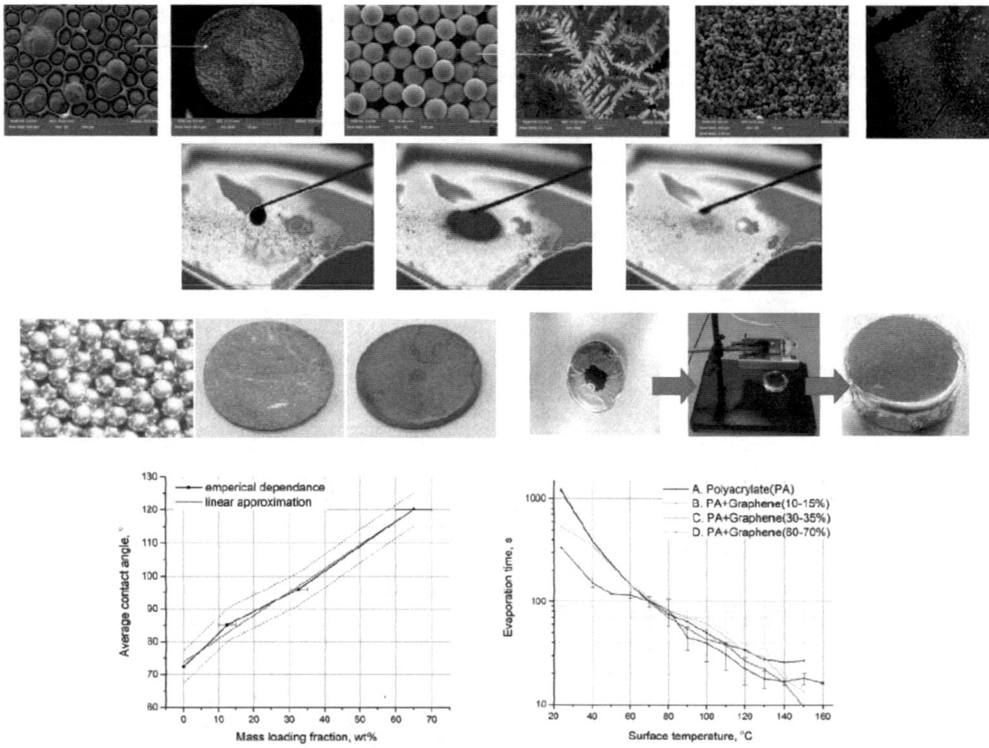

Figure 3. Research results of undergraduate and graduate students in the framework of educational programs.

2. Machine learning methods in educational curriculum (training program)

2.1. Content and learning outcomes

Most of the subjects of the program are courses that include not only lectures, labs and workshops, but also specialized master's programs in machine learning. Our master's program in machine learning consists of several sections, which include: the development of algorithms for finding patterns in the analysis of experimental data, the development of computer models for processing digital data obtained during experimental research (data of optical, Raman or electron microscopy, etc.), systems for presenting digital data in the form of visual images, analysis of graphical patterns by statistical methods, including a numerical assessment of the reliability and uncertainty intervals of the obtained big data. The program consists of two sections: the first - general courses on artificial intelligence and machine learning, the second - specialized courses and practical exercises in the sections: methods and models of machine learning in nanomaterials and nanotechnologies, analysis and synthesis of big data (deterministic and statistical methods and algorithms) in nanotechnology and nanomaterials. In addition, practical seminars are held at which master's or doctoral students pose and solve specific problems using machine learning methods in the field of spectra of electronic and phonon states in low-dimensional structures, analyze the processes of wetting and spreading of drops, processes of evaporation and boiling on nanostructured substrates, using large databases. both existing experiments and databases and knowledge obtained directly during the implementation of research, bachelor's or master's theses.

The content of courses and disciplines based on machine learning may contain:

- methods of analysis based on statistical data processing;

- methods of mathematical modeling based on the theory of hierarchical systems and supercomplex systems, including new computer algorithms;
- the influence of the dimensional classical and quantum effects on the obtained experimental results;
- methods for analyzing the results of observations, for example, in the field of wetting and spreading of liquids on nanostructured surfaces, where there are a huge number of factors affecting the final results;
- special approaches related to graph theory;
- new methods of visualization of observations and algorithms for processing electron and optical microscopy, using the latest approaches in comparative analysis and synthesis of digital images, etc.

In various fields of nanotechnology, the amount of data is growing every year, as well as the complexity and depth of their processing. This applies to nanoplasmonics, promising nanomaterials, nanoelectronics and optoelectronics. At the same time, the volume of these obtained data is interconnected in a very complex way. These connections are often not obvious and require a special method of analysis and synthesis. One of these approaches is today the machine learning method, which allows you to perform a number of functions:
- development of training models that allow you to process, analyze and synthesize large datasets;
- to classify and combine the results of observations into families according to certain criteria;
- to determine the functional properties of certain families, as well as to introduce the efficiency of certain processes, systems or elements;
- to predict experiments according to certain rules;
- check the hypotheses being introduced and, using certain algorithms, determine their reliability or uncertainty.

The hierarchical structure of the interaction between machine learning and nanoscience is shown in Figure 4.

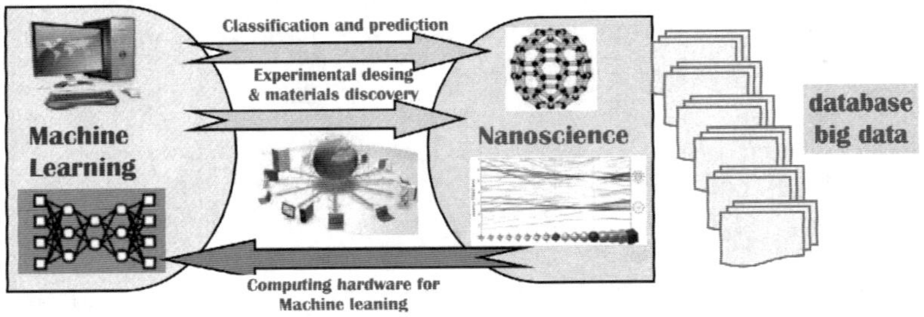

Figure 4. Hierarchical structure of interaction between nanoscale science and machine learning systems.

In nanoscience, high-throughput experiments, small sample sizes, fast and high-resolution instruments (optical and electron microscopy), and correct imaging are becoming increasingly important.

For example, in nanophotonics, physics and technology of new nanomaterials and nanoplasmonics the properties of objects change systematically, the amount of data is constantly growing, and the complexity of their processing is also growing. Increasingly sophisticated research methods are generating datasets that are too vast and complex for researchers to mentally understand without computational assistance; however, this data is rich in relationships that researchers would like to understand. Machine learning allows researchers to:
- analyze large datasets using training models that can be used to classify observations into discrete groups, find out which features determine the performance indicator, or predict the results of new experiments;
- even in areas where large data sets are not typical, help researchers design experiments to optimize performance or test hypotheses more effectively;

2.2. Machine learning results

We have introduced the criteria for the final stage of machine learning, a set of competencies and specific knowledge, which testify to the positive assimilation of the course:
- knowledge and ability to formulate models and methods of machine learning;
- knowledge of various examples of machine learning in various sections of nanotechnology and nanomaterials;
- the ability to use models and algorithms for more efficient analysis of big data;
- the ability to use computer algorithms and computational models for comparative analysis of data and their synthesis according to certain criteria;
- understand the relationship between artificial intelligence and machine learning methods;
- the ability to build machine learning methods for specific sections of nanotechnology and nanomaterials and substantiate such methods.

We have developed a new algorithm for studying the intersection of machine learning and nanoscience, which in practice of working with students and graduate students showed that several types of their mutual influence can be distinguished:
- the use of machine learning methods and tools for the analysis of various physical properties of a large family of nanomaterials (for example, analysis and synthesis of data on nanocomposites for materials with a thermal interface);
- the discovery of new types of nanomaterials using machine learning to plan new experiments (for example, when choosing nanofluids for solar thermal energy);
- analysis of nanoscientific methods and properties of nanomaterials to expand the capabilities of machine learning through the development of new equipment (for example, new sensors based on graphene or new types of lithium-ion batteries).

We are confident that the synthesis of machine learning and nanoscience, the development of new algorithms for the integration of various approaches and the widespread use of similar methods will significantly expand the opportunities for learning about the program "nanotechnology and nanomaterials in energy"

Acknowledgments
The work is supported by the Russian Science Foundation (Grant No. 17-19-01757).

References

[1] Dmitriev A and Klimenko 2020 Heat power engineering **2**, 1-16
[2] A.S Dmitriev. Hybrid Graphene Nanocomposites: Thermal Interface Materials and Functional Energy Materials In: Graphene Production and Application // Ed. Prof. Sadia Ameen. 2019. InTechOpen.
[3] Kluender E, Hedrick J, Brown K et al 2019 Proc. Natl. Acad. Sci. U. S. A. **116(1)** 40-45
[4] Bzdok D, Altman N, Krzywinski M Nat. Methods 2018 **15** 233−23
[5] Cao B, Adutwum L, Oliynyk A et al 2018ACS Nano **12 (8)** 7434−7444
[6] Ma W, Cheng F, Liu 2018 ACS Nano**12 (6)** 6326−6334
[7] Carrasquilla J, Melko R Nat. Phys. 2017, **13 (5)** 431
[8] Carpenter E, Stanger B, Issadore D 2017 ACS Nano **11(11)** 11182−11193
[9] Rashidi M, Wolkow R 2018 ACS Nano **12 (6)** 5185−5189

Integrated development of low-temperature thermal waters

D A Alkhasova and B A Alkhasov

Institute for Geothermal Research and Renewable Energy of Joint Institute for High Temperatures of the Russian Academy of Sciences, Makhachkala, Russia

alkhasova.dzhamilya@mail.ru

Abstract. The research concerns the efficient development of low-potential thermal waters in the East-Ciscaucasian artesian basin. A technology is proposed for the integrated development of low-temperature thermal water, using its heat potential for district heating and hot water supply, as well the water itself for various water management purposes, with its quality previously brought to the standards of drinking water. Various chemical water treatment units are suggested for this purpose, with their design and technological features being formed depending on the quality of the source water. The system will enable the maximization of the resource potential of the geothermal well and its all-year-round operation. The paper shows the efficiency of using the potential of geothermal energy resources in energy- biological complexes. The geothermal-biogas technology with the integrated application of thermal waters for various needs is presented. Such thermal water utilization provides for the most efficient use of its thermal potential with a decrease in temperature to a value close to the ambient temperature.

1. Introduction

In the East Ciscaucasian artesian basin (ECAB) with the area of more than 200 thousand km^2, huge resources of low-temperature thermal waters are bedded at depths of 200 - 700 m. They are used to supply water to settlements and various objects. Low-temperature waters are pressure overflowing waters with an ejection height from 10 to 120 m. The deeper the aquifer lies, the higher the temperature of the extracted water (25 - 60 ° C) and, as a rule, the height of the overflow. Predicted production resources with an average temperature of 40° C are 1.5 million m^3 / day [1]. Within the basin there are numerous wells drilled for overflowing water with temperatures of 50 - 60° C and salinity of 1 - 3 g / dm^3. These resources are promising for integrated development with utilization of water and heat potentials.

The productive use of the energy potential of geothermal resources can be achieved through their comprehensive development, but at present their potential is not fully used. There are great prospects for the construction of integrated systems of geothermal energy supply at most of the thermal water deposits in the Russian Federation.

Figure 1 shows the geothermal map of ECAB [2].

Content from this work may be used under the terms of the Creative Commons Attribution 3.0 licence. Any further distribution of this work must maintain attribution to the author(s) and the title of the work, journal citation and DOI.
Published under licence by IOP Publishing Ltd

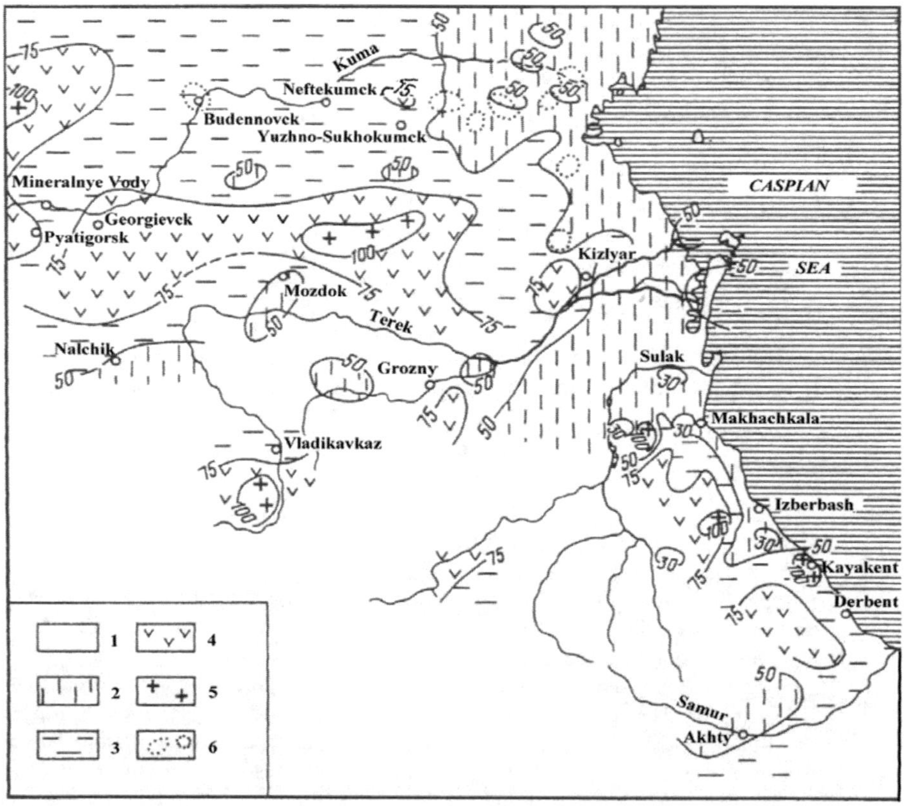

Figure 1. Geothermal map of East-Ciscaucasian artesian basin.
Values of heat flux, mW/m^2: 1-<30 are the negative anomalies on heat flux; 2-30 – 50; 3-50 – 75; 4-75 -100; 5- >100; 6-thermoanomalies of the bedding

In previously published works [3,4], technologies were proposed for the integrated development of geothermal resources of various potentials with reference to specific geothermal fields. Technological solutions were offered for the integrated development of high-temperature hydro-geothermal resources of the North Caucasus region with the conversion of thermal energy into electricity in a binary GeoPP and subsequent extraction of dissolved chemical compounds. An assessment of the efficiency of the proposed technologies was made using physical and mathematical, thermodynamic and optimization methods of calculation, and physical and chemical experimental investigations. Studies were carried out to protect the power equipment of geothermal systems from hard scaling. To date significant progress has been reached in the production of geothermal energy in the world [5-7].

2. Technologies for the integrated development of low-temperature thermal waters

High economic efficiency of low-potential thermal waters can be achieved by their integrated development using the thermal potential for energy needs, and the water itself for various water management purposes. The technological scheme shown in Figure 2 demonstrates the possibilities of such options of water utilization.

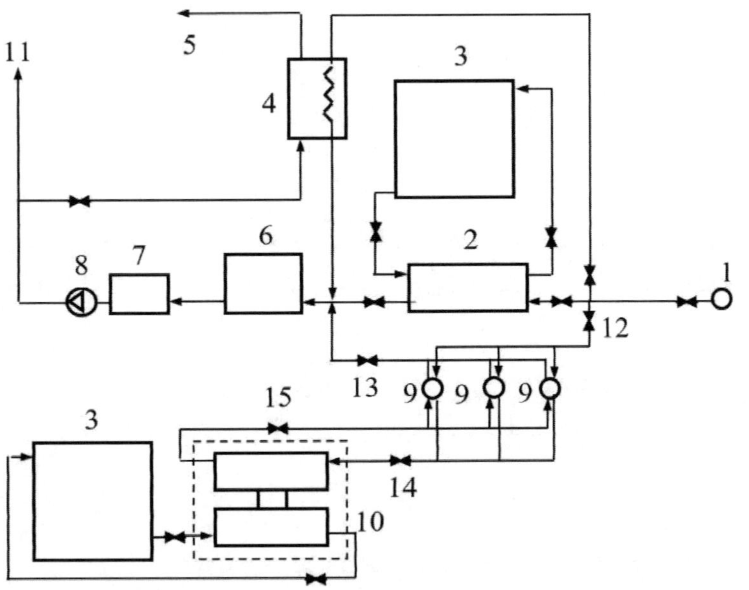

Figure 2. The system of geothermal heat and water supply.
1 geothermal well; **2** heat exchanger; **3** low-temperature floor heating system; **4** thermo-insulated tank; **5** to hot water supply; **6** chemical water treatment unit; **7** reservoir of pure water; **8** pumping station; **9** heat storage wells; **10** heat pump; **11** to cold water supply; **12**; **13**; **14**; **15** valves

During the heating period, the thermal potential of water is used in the low-temperature floor heating system and for heating water in the hot water supply system. The water cooled in the heat exchangers enters the chemical water treatment unit, and then goes to consumers. In the inter-heating period, part of the thermal water from the well, used in the floor heating system, enters the down-hole heat exchangers to restore the heat field around them, and the water cooled in the wells goes to the chemical water treatment units. During the heating season, the heat accumulated in the rock is used in the floor heating system with a heat pump.

When developing geothermal heat supply systems, it is necessary to ensure the maximum value of the utilization efficiency rate of thermal water intake η_{geot}. It is the ratio of the well's actual thermal potential used during the year to the maximum amount of heat that can be obtained during year-round operation of the well at a flow rate corresponding to operating reserves and temperature drop of the used water up to conditional temperature. The value of the coefficient η varies in the range: heating 0.05 - 0.34; and hot water supply 0.70 - 0.92. The given system will maximize the use of the resource potential of a thermal well and provide its year-round operation with the approximation of η_{geot} to the factor for hot water supply.

Figure 3 shows a geothermal-biogas technology with the integrated use of thermal waters. In such a system, part of the low-temperature thermal water (55 - 60 ° C) from the well goes to the heat exchanger, where it heats the raw material in the methane tank. Another part of water goes to the greenhouse complex. The required temperature regime in the methane tank is maintained by heated water circulating in tubular coils laid on the bottom of the methane tank. From the heat exchanger, thermal water is sent to the subsurface heating system and then, at a lower temperature, enters the fish breeding pool, which also receives another part of water from the greenhouse complex. Such an integrated use of thermal water for various needs provides for the most efficient realization of the thermal potential of geothermal water, with its final temperature being close to the ambient temperature.

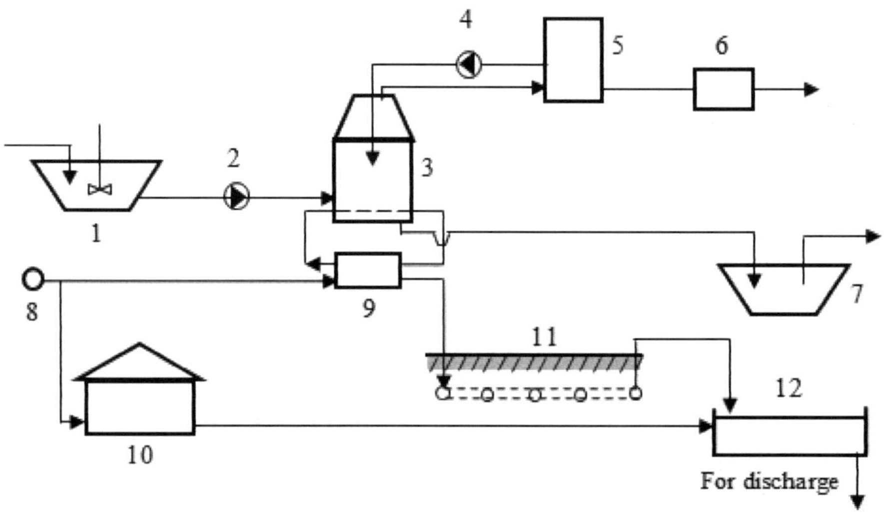

Figure 3. Technology of integrated use of thermal water.
1 container for the preparation of raw materials; **2** dosing pump; **3** methane tank; **4** compressor; **5** gas holder; **6** biogas purification devices; **7** fertilizer collection; **8** geothermal well; **9** heat exchanger; **10** greenhouse complex; **11** ground heating system; **12** fish breeding pool

Energy-biological complexes are promising objects for the integrated utilization of geothermal resources.

3. Conclusion

The widespread introduction of technologies for the integrated development of low-temperature thermal waters will largely solve the problem that is relevant for consumers of the North Caucasus region: providing them with drinking water of high quality. According to various estimates, the number of wells in the region for the extraction of low-temperature water ranges from 7,000 to 10,000. The water salinity of most wells does not exceed 1–3 g / l. However, many of them have been decommissioned by now for various reasons.

The Institute for Geothermal Research and Renewable Energy of the JIHT RAS (Dagestan, South of Russia) conducts long-term field studies of the low-potential thermal waters of the East Ciscaucasia artesian basin. The waters have been typified and pollutants have been established: arsenic compounds are determined as the most toxic of them. The composition of organic components in thermal waters has been studied in details: volatile phenols and humus acids are found to be the most dangerous. Effective technologies have been developed for the purification of low-potential thermal waters from arsenic and organic pollutants.

References

[1] Alkhasov A B, Alishaev MG, Alkhasova D A, Kaymarazov A G and Ramazanov M M 2012 *Development of low-potential geothermal heat* (Moscow: Fizmatlit) p 280

[2] Kurbanov M K 2001 *Geothermal and Hydromineral Resources of the Eastern Caucasus and Ciscaucasia* (Moscow: Nauka) p 260

[3] Alkhasov A B, Alkhasova D A, Ramazanov A Sh and Kasparova M A 2017 *Therm. Eng.* **64**(9) 637–643 DOI: 10.1134/S0040601517090014

[4] Alkhasov A B and Alkhasova D A 2019 *Therm. Eng.* **66**(5) 361–366 DOI: 10.1134/S004060151905001X

[5] Gharibi S , Mortezazadeh E, Hashemi Aghcheh Bodi SJ and Vatani A 2018 *Energy* **153** 554–567. DOI:10.1016/j.energy.2018.04.003

[6] Wei Zhang, Zhanqing Qu, Tiankui Guo and Zhiyuan Wang 2019 *Renew Energy* **143** 855-871. DOI: 10.1016/j.renene.2019.05.054
[7] Gaosheng Wang, Xianzhi Song, Yu Shi,RuiyueYang, FeixueYulong, Rui Zheng and Jiacheng Li 2021 *Renew Energy* **163** 974-986. DOI: 10.1016/j.renene.2020.08.121

Improving a heat supply system by increasing the efficiency of the heat pump unit

P V Rotov and R A Gafurov

Ulyanovsk State Technical University, ul. Severnyi Venets 32, Ulyanovsk, 432027 Russia

gafurus@mail.ru

Abstract. The reliability and efficiency of the operation of district heating systems is largely determined by the efficiency of preparation of heating network water. In open heat supply systems, make-up water, among other things, compensates for the water intake in hot water supply systems. A number of technologies have been developed that increase the efficiency of an open heat supply system by reducing the water consumption in the supply pipeline of the heating network, increasing the operating time of the heat pump, and increasing the specific generation of electricity for heat consumption at the CHP plant due to additional cooling of the network water in the return pipe of the heating network.

District heating systems are a set of interconnected devices and equipment that cover most of the heat load of cities and towns. Most domestic heat supply systems operate according to a high-quality heat schedule. Coolant is prepared at a CHP plant, regulated depending on the outside air temperature, and supplied through a pipeline system to the heating and hot water supply systems of to consumers.

Due to the intensive regulation of heat consumption by end users, the efficiency of the heat source can be significantly reduced. The use of heat pumps at intermediate points of the heat supply system makes it possible to compensate for the influence of the operating modes of the heat consumption systems of end users on the efficiency of the thermal power plant.

Scientific and technical studies have shown that low-grade heat sources can be an effective complement to the traditional heat supply system. The heat of return network water can be used as one of such sources of heat [1, 2]. The use of a heat pump installation allows independent heating of water for hot water supply, which makes it possible, first, to provide backup heat supply, if necessary, and second, to increase the radius of effective operation of the heat supply system in cases of insufficient capacity of pipelines of heating networks to remote consumers.

A significant contribution to the development and justification of the use of heat pumps in centralized heat supply systems has been made by the staff of the laboratory of Heat Power Systems and Installations (headed by Prof. V I Sharapov) at the Ulyanovsk State Technical University (USTU [3-6].

A special feature of the developed technologies is the heating of heat network water to the required temperature using a heat pump.

Major advantages of these technologies are reduced cost of heat transport, reduced heat loss in the heat supply system, reduced fuel consumption, and increased electricity generation based on heat consumption.

Content from this work may be used under the terms of the Creative Commons Attribution 3.0 licence. Any further distribution of this work must maintain attribution to the author(s) and the title of the work, journal citation and DOI.
Published under licence by IOP Publishing Ltd

At the same time, some operating restrictions reduce the energy-saving potential of the technologies. One of these restrictions is the use of return network water only in the transition range of the temperature schedule of the central quality control without a lower break.

A method for improving the operation efficiency of an open heat supply system is proposed. The improved heat supply system is shown schematically in Fig. 1.

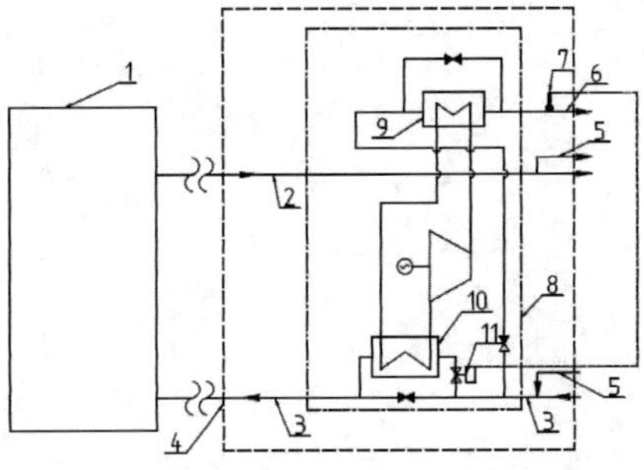

Figure 1. Open heat supply system: 1 is a heat source; 2 is a supply pipeline; 3 is a return pipeline; 4 is a heat point; 5 are pipelines of the heating system; 6 is a hot water system; 7 is a temperature sensor; 8 is a heat pump; 9 is a condenser; 10 is a vaporizer; 11 is a temperature controller.

A distinctive feature of the new solution is that analysis of water for hot water supply is carried out only for the return pipeline of the heating network.

To determine the effectiveness of this solution, we analyzed the operating modes of the steam compression heat pump installation in the heat supply system for different initial data. Figure 2 schematically shows the heat pump design for which the parameters of the operating points of the heat pump cycle were determined using the Solkane 6.0 software. Ozone-safe R-134a refrigerant is used as the working fluid for the heat pump. The low-grade source of heat is return network water with a temperature of 35–50 °C. In the heat pump, the water for hot water supply is heated to a temperature of 65–70 °C. In the calculation of the heat pump cycle, the electromechanical efficiency of the pump was set equal to 0.95 and the adiabatic efficiency equal to 0.8.

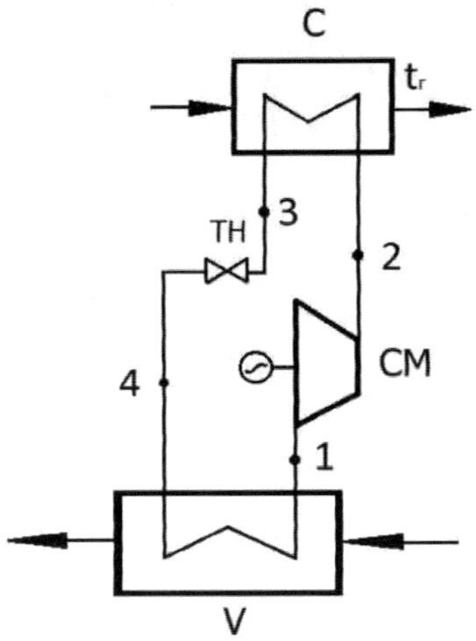

Figure 2. Schematic design of the heat pump: C is a condenser; V is a vaporizer; CM is a compressor; TH is a throttle; 1, 2, 3, and 4 are characteristic points of the cycle.

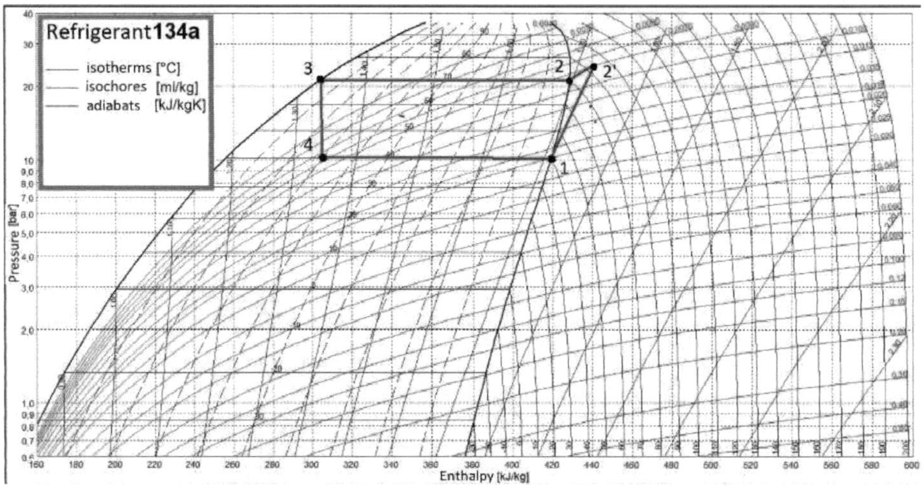

Figure 3. Operating cycle of the heat pump

The analysis and calculation of the heat pump cycles under the specified conditions confirmed the possibility of a stable temperature regime of the refrigerant at points 2 and 3 with a variable refrigerant temperature at point 1.

Figure 3 shows a diagram of the heat pump operation cycle for the following initial data: temperature of the low-grade heat source t_{gs}=45°C; temperature of the refrigerant at the outlet of the evaporator t_1=40°C.

The efficiency of the developed technology is calculated by the method of specific electricity generation based on thermal consumption [7, 8]. The efficiency of the new technology is estimated by the amount of annual savings of conventional fuel due to an increase in electricity generation based on thermal consumption with an additional decrease in the temperature of the return network water in the heat pump installation.

The initial data for the calculation are: steam capacity of the boiler 500 t/h; temperature of the network water supplied to the consumer after the network heaters t_1 = 77 °C; return water temperature in normal mode t_2 = 45 °C; return water temperature taking into account the operation of the heat pump installation t_{2new} = 40 °C; cost of a ton of conventional fuel 4400 rubles. The calculation used the operating parameters of the turbo stop given in [7].

The thermal power developed by the turbine based on heat consumption due to the upper and lower network selections in the traditional mode N_{tf}^{trad}, MW, is:

$$N_{tf}^{trad} = (D_{unh}^{trad} \cdot (i_s - i_{unh}) \cdot \eta_{em}) + (D_{lnh}^{trad} \cdot (i_s - i_{lnh}) \cdot \eta_{em}) \quad (1)$$

where is, i_{unh}, and i_{lnh} are the enthalpies of the hot steam, thge pair of upper and lower heaters for heating the network, and make-up water flows; D_{uhn}^{trad} and D_{lnh}^{trad} are the consumptions of steam released from the selections for heating the network water in the upper and lower heaters;, η_eм is the electromechanical efficiency.

The thermal power developed by the turbine based on heat consumption due to the upper and lower network selections in the new mode $N_{tf}^{new.}$, MW, is

$$N_{tf}^{new} = (D_{unh}^{new} \cdot (i_s - i_{unh}) \cdot \eta_{em}) + (D_{lnh}^{new} \cdot (i_s - i_{lnh}) \cdot \eta_{em}) \quad (2)$$

The difference in the thermal power developed by the turbine based on heat consumption between the traditional and new methods ΔN_{tf}, **MW**, is

$$\Delta N_{tf} = N_{tf}^{trad} - N_{tf}^{new} \quad (3)$$

The annual savings of conventional fuel from the use of the new solution are

$$\Delta B = (\Delta N_{tf} \cdot n - N_{comp.}) \cdot \Delta b_e \quad (4)$$

where Δb_e is the difference in the specific fuel consumption for electricity generation for the condensation and heating cycles; N_{comp} is the the power developed by the compressor of the heat pump installation; n is number of days of the heating period.

The savings in monetary terms E, rub, are

$$E = \Delta B \cdot P \quad (5)$$

where P is the cost of conditional fuel.

Calculation The annual saving of conventional fuel calculated using equations (1–5) is 3425 tons (14.7 million rubles per year), which allows reducing the payback period of the heat pump by at least 2 years compared to previously developed technologies [5, 6].

Technical and economic calculations confirm the sufficient efficiency of the proposed technology.

Conclusions

1. One of the ways to increase the efficiency of centralized heat supply systems is to include in them heat pump installations that use low-grade heat sources and, in particular, the potential of return network water.
2. Technology of hot water preparation using a heat pump that allows the use of a wider temperature range of return network water has been proposed and technically and economically justified.
3. The developed technical solution allows an increase in the life of the heat pump installation, a reduction in water consumption in the supply pipeline of the heat network, an increase in the specific generation of electric energy based on heat consumption at a CHP due to additional cooling of network water in the return pipeline of the heating network, and, hence, a reduction in the payback period of the proposed technology.
4. The annual savings from the use of the proposed technology are more than 14.7 million rubles.

References

[1] Andryushchenko A I, Possible fuel economy from the use of recycling TNUS in the power supply system of enterprises. - Industrial Power Engineering, 2003, No. 2

[2] Nikolaev Yu E Baksheev A Yu Determination of the efficiency of heat pumps using the heat of the return network water. - Industrial Power Engineering, 2007, No. 9

[3] Rotov P V, Sharapov V I Technical and economic assessment of the use of heat pump installations in centralized heat supply systems // Industrial energy. 2015. No. 6. pp. 6-11.

[4] Rotov P V, Orlov M E, Sharapov V I, Sivukhin A A Improving the efficiency of centralized heat supply systems through the use of heat pump installations / / Industrial power engineering. 2014. No. 7. pp. 27-31.

[5] Rotov P V and Sharapov V I 2013 Regulation of loads on municipal district heating systems textbook (Ulyanovsk: USTU) p 309

[6] Patent 2474765 (RU). MPC7 F24D 3/00. Method of operation of an open heat supply system / P V Rotov, M E Orlov, V I Sharapov, P E Chaukin, V E Mordovin / / Bulletin of Inventions. 2013. №4.

[7] Sharapov V I, Pazushkin P P, Tsyura D V, Makarova E V. Calculation of the energy efficiency of water treatment technologies at a thermal power plant: A textbook/ - Ulyanovsk: USTU, 2003. - 120 p.

[8] Zamaleev M M, Sharapov V I, Gubin I V, Pavlov V A Feasibility study of new technologies for snow utilization at thermal power plants. News of higher educational institutions. ENERGY PROBLEMS. 2016. No. 11-12. p. 3-9.

Circulation management in hot water supply systems

P V Rotov[1], A A Sivukhin[2], M A Rotova[1], R A Gafurov[1] and A V Gorshkov[3]

[1] Ulyanovsk State Technical University, Ulyanovsk, Russia
[2] Ulyanovsk Municipal Unitary Enterprise "Municipal Heat Service", Ulyanovsk, Russia
[3] Ulyanovsk Branch of the T Plus PJSC, Ulyanovsk, Russia

p.rotov@rambler.ru

Abstract. The compliance of the actual heat consumption for heating cold water for public hot water supply service with the approved standard has been analyzed. Using passive engineering experiments, we studied the operating modes of hot water supply systems for several groups of houses, in which different methods for regulating water consumption in hot water supply systems were used. Data were collected using online energy monitoring and commercial metering system. Features of the static and dynamic load regulation of hot water systems have been investigated. The effect of methods for regulating the heat load of hot water supply systems on their actual heat consumption has been analyzed. An assessment has been made of the possibility of bringing the actual operating mode of the hot water systems to the specified value which is the basis for calculating the consumption rate. Ways to improve the efficiency of hot water supply systems have been determined. Conclusions have been drawn concerning the need to revise the standardized indicators of hot water quality and adjust the methodology for calculating the consumption rate in hot water supply systems.

One of the key performance indicators of district heating systems is the return water temperature, which is determined by the accuracy and quality of load regulation in heating, ventilation, and hot water supply systems [1, 2].

The regulation of return water temperature in heating systems and the absence of regulation of circulating water temperature in hot water systems lead to an increased influence of the operating mode of hot water systems on the temperature of network return water [3].

At present, the consumption in hot water supply systems is calculated using the standardized heat consumption of for heating cold water, which in the regions of Russia varies from 0.057 to 0.069 Gcal/m^3 [4, 5]. Studies have shown that the heat consumption rate for water heating is widespread and significantly underestimated. In re-circulation systems with a hot water temperature between 60 °C and 70 °C, the actual specific heat consumption for hot water supply can exceed 0.1 Gcal/m^3, which leads to significant annual financial losses of energy utilities, which, according to experts, can reach 100 million rubles for a city with a population of around 700 000 people.

A number of heat consumption systems in the Ulyanovskheat supply system were surveyed to determine whether the actual and normative heat consumptions for heating cold water are consistent. The operation of hot water supply systems was compared with the traditional circulation flow limitation using constant-section orifice valves and by the installation of thermostatic valves.

Content from this work may be used under the terms of the Creative Commons Attribution 3.0 licence. Any further distribution of this work must maintain attribution to the author(s) and the title of the work, journal citation and DOI.
Published under licence by IOP Publishing Ltd

The evolution of the specific heat consumption before and after technical measures in residential buildings connected to one of the central heat exchange stations is shown in Figure 1.

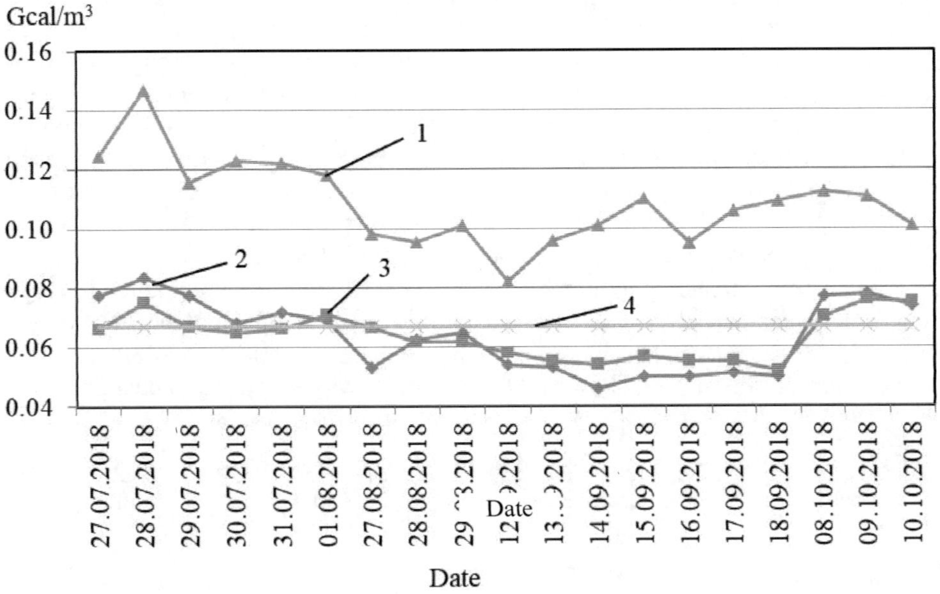

Figure 1. Dynamics of specific heat consumption in the systems of the central heat exchange station of the Locomotivnoe Depo central heat supply station: 1 – ul. Zheleznodorozhnaya 45; 2 – ul. Kirova 50; 3 – ul. Kirova 52.; 4 – estimated heat consumption rate for heating water for hot water supply.

The following results were obtained from the technical measures carried out in the residential buildings:

1. Prior to the above technical measures, the actual consumption in the hot water supply systems was significantly (more than twice) higher than the approved normative consumption of heat energy for heating cold water for hot water supply service. The reason for the discrepancy is the design features of hot water systems due to which the heat energy from the internal hot water supply system is used to heatbathrooms irrespective of the time of day and outside temperature.

2. The specific heat consumption is significantly influenced by the hot water consumption mode. The specific heat consumption in circulation systems decreases to 0.054 Gcal/m^3 during peak hours and rises to 0.140–0.20 Gcal/m^3. In non-circulating hot water supply systems, the influence of the amount of drained coolant on the actual specific heat consumption is not as significant as in circulating systems. During peak and minimum hours, the heating coefficient varies between 0.053 and 0.081 Gcal/m^3. Obviously, a greater effect can be achieved by using dynamic control methods that take into account the irregularities in the operation of hot water systems [6–8]

3. The actual specific heat consumption was brought to the approved normative values of 0.062 Gcal/m^3 without compromising the quality of the end user's hot water supply, only in houses not equipped with towel heaters.

4. The technical measures implemented to limitthe circulation flow or maintain the set value of the circulating water temperature do not fully bring the actual specific heat consumption to the normative values without deteriorating the quality parameters. It can be assumed that when lowering the lower limit of the permissible range of hot water temperature control, for example, to 55° C, the efficiency of such technical measures will be higher and will not affect the sanitary reliability of hot water supply systems even without the use of corrective measures.

It has been established that restrictive measures alone cannot ensure that the actual operation of hot water supply systems meets the approved normative values. One way to solve this problem could be to dynamically adjust the hot water supply load according to the consumption mode. Technically, this is solved by installing control equipment on the circulation line of the hot water supply system. The equipment is adjusted so as to reduce the circulation flow rate during peak demand and compensate for heat loss in the hot water system by draining water. During the hours of minimum water withdrawal, losses in the hot water system are compensated for by increasing the circulating water flow.

In 2019–2020, 46 control units were installed in 28 buildings supplied by the Ulyanovsk heat supply system. The main equipment of the control units included TA-Modulator combined control valves with a TA-Slider 160 electric actuator, OWEN DTS3225-PT1000.B2 surface-mounted water temperature sensors with a Pt1000 sensitive element, and OWEN-PR200 controllers with freely programmable logic.

Initially, the control units were set to maintain the temperature in the return pipeline within -50– 52 °C around the clock. Decreasingthe temperature by 2–3 °C did not have a tangible effect. Therefore, the control units were partially readjusted according to the actual hot water supply consumption schedule (Table 1). The calculated daily average water temperature in the circulation line was 44 °C. After applying the temperature schedule shown in Table 1, there was a significant change in the operation of hot water systems.

Table 1. Temperature control schedule for temperature control in the circulation line

Hours	Temperature, °C	Hours	Temperature, °C
00:00	48	12:00	45
01:00	46	13:00	45
02:00	45	14:00	45
03:00	44	15:00	45
04:00	44	16:00	45
05:00	45	17:00	45
06:00	48	18:00	40
07:00	48	19:00	40
08:00	45	20:00	40
09:00	40	21:00	40
10:00	40	22:00	48
11:00	40	23:00	48

The study showed that the actual average daily water temperature in the circulation pipeline decreased by more than 6 °C. The average daily water consumption in the hot water supply and circulation pipelines decreased by 37% and 48%, respectively, which resulted in a corresponding reduction of heat consumption by the system. The average reduction in heat consumption for water heating was more than 15%.

The average savings per control unit was more than 90 Gcal/year. It can be assumed that the total annual savings for all 46 units will be more than 2.9 million rubles. With a cost of construction and installation work of 3.5 million rubles, the simple payback period will be 1.2 years.

However, simultaneously with the positive economic effect, there was a decrease in the water temperature in the hot water supply system during the hours of minimum consumption, which was due to the reduced circulation flow at these times. Thus, in the periods from 00:00 a.m. to 4:00 a.m. and from 10:00 a.m. to 8:00 p.m., the water temperature in the DHW supply pipeline was below 60°C. This deviation required an additional readjustment of the control units and an increase in the water temperature in the hot water supply pipeline. The average daily water temperature in the circulation line after readjustment was 44 °C.

A change in the temperature of the return network water results in a proportional change in the production of electricity for the heat demand [9, 10]. According to [9], the change in the turbine capacity $\Delta N_{\text{тур}}$, kW/MW, per 1 MW of the heat load of the turbine can be defined as

$$\Delta N_{\text{тур}} = 4\Delta t_{\text{ос}}\alpha\eta_{\text{эм}} \qquad (1)$$

where $\Delta t_{\text{ос}} = t_{\text{ос}}^{\text{баз}} - t_{\text{ос}}^{\text{изм}}$ is the difference between the base $t_{\text{ос}}^{\text{баз}}$ and modified $t_{\text{ос}}^{\text{изм}}$ return water temperatures,°C; α is the thermodynamic coefficient taking into account the relationship between the saturation temperature and the heat drop at the points of adiabatic steam expansion in the turbine, kW/(kg·°C); $\eta_{\text{эм}}$ is the electromechanical efficiency of the turbine generator.

Calculations using formula (1) show that at $\Delta t_{\text{ос}}$ = 4–6°C, the electricity generation using heat consumption $\Delta N_{\text{тур}}$ increases by 13–20 kW per 1 MW of heat load from turbine extraction.

Studies have shown that consumption-based flow control in the circulation line can optimize the operation of the hot water supply system and provide significant technical and economic benefits. However, dynamic load control in the hot water supply system makes it impossible to achieve the design standard for heating and to achieve a significant drop in water temperature after the hot water system. With an average daily water temperature in the circulation pipeline of 44–46°C, there are already modes that do not provide the standard quality of hot water. It is safe to say that during the transitional hours of the heating period, when the water temperature in the heating systems is significantly lower than 44 °C, the operating mode of hot water supply systems will negatively affect the energy efficiency of the thermal power plant. A way out of this situation may be to reduce the standard temperature of hot water to 55 °C and to expand the range of its regulation [11] or to revise the methodology for establishing a standard value for the consumption of thermal energy for heating water or reject the use of this standard value. This will not deterioratethe sanitary reliability of the systems and will provide an opportunity to realize the hidden energy saving potential in hot water systems.

Conclusions

The calculated standardized consumption of thermal energy for heating cold water does not correspond to the actual operating conditions of hot water supply systems. This is due to the need to maintain a higher hot water temperature in the heat supply system to meet the quality standards of end customers and to maintain excessive circulation flow to compensate for the heat losses during periods of minimum water withdrawal.

The well-known and widely used static control methods for limiting the circulation flow by using throttling valves or by installing thermostatic valves do not allow the actual heat consumption to be brought close to the standardized values and can lead to violations of the current hot water quality.

The efficiency of hot water systems can be significantly improved by dynamically adjusting the flow and temperature of the circulating water to account for daily and weekly irregularities in their operation.

In the Ulyanovskheat supply system, dynamic load regulation of the hot water system was implemented for a number of residential buildings. Despite the proven efficiency, this control method did not provide the design value of the heat consumption standard for hot water supply heating, and a significant reduction in the temperature in the circulation line was impossible due to the regulatory requirements for the quality of hot water at the water draw-off points.

To increase the efficiency of the control methods used, the range of possible regulation hot water temperature should be legally expanded. A short-term decrease in hot water temperature to 50–55°C will not affect the sanitary reliability of the hot water supply systems.

In our opinion, the existing method for calculating the consumption in hot water systems does not allow the actual heat consumption to be determined. It is advisable to revise the methodology for standardizing heat consumption for water heating or abandon the normative method of calculation.

References

[1] Rafalskaya T A 2019 Investigation of the possibility of organizing low-temperature heat supply with central quality control *Therm Eng* **11** 102-12.

[2] Xiaofang S, Peng W, Panhong R and Hua Z 2016 The Influence of Central Regulation Methods upon Annual Heat Loss in Heating Network *MATEC Web of Confer.* **54** 06004 DOI: 10.1051/matecconf/20165406004/

[3] Rafalskaya T A and Mansurov R Sh 2017 Assessment of the influence of water temperature in the hot water supply system on the temperature regime of premises *Water supp. and sanitary equipm.* **4** 42-9.

[4] Semenov V G 2018 Standards for Hot Water Heating *Heat Supply News* **6-7** 8-17.

[5] Rotov P V, Sivukhin A A, Gafurov R A and Rotova M A 2020 About standards for hot water heating *J. of Phys.: Conference Series* **1683** 042017.

[6] Rotov P V, Zhukov D A, Zhukova A V and Sivukhin A A 2017 About economy of fuel and resources in the hot water supply system *J. of Phys.: Conference Series* **891** 012160.

[7] Rotov P V and Sivukhin A A 2016 Evaluation of efficiency of hot water supply load control technologies *Energy saving and water treatment* **6** 22-2.

[8] Tumanova T and Cimbale A 2015 The Technical-Economic Analysis of Hot Water Supply Systems for Residential Buildings *Advanced HVAC and Natural* Gas *Technologies* 177–83 DOI: 10.7250/rehvaconf.2015.025.

[9] Yakovlev B V 2008 *Improving the efficiency of heating and heat supply systems* (Moscow: Heat supply news) p 448.

[10] Zamaleev M M, Sharapov V I, Gubin I V and Pavlov V A 2016 Feasibility study of new technologies for snow utilization at CHPP *Power engineer.: resear.,equip., technol.* **11-12** 3–9.

[11] Pieper H, Ommen T S, Markussen W B and Elmegaard B 2017 Optimal Usage of Low Temperature Sources to Supply District Heating by Heat Pumps *Proc. of ECOS 2017: 30th International Conference of Efficiency, Cost, Optimization, Simulation and Environmental Impact of Energy Systems* (San Diego: California, USA, 2–6 July 2017) http://orbit.dtu.dk/files/134463914/ECOS_2017_paper_191.pdf.

AUTHOR INDEX